Donald S. McKenzie

INTERNATIONAL SERIES IN
NATURAL PHILOSOPHY
GENERAL EDITOR: D. TER HAAR

VOLUME 108

THEORY OF PHASE TRANSITIONS: RIGOROUS RESULTS

THEORY OF PHASE TRANSITIONS: RIGOROUS RESULTS

By

YA. G. SINAI

D. Sc. (Math. Phys.)
Landau Institute of Theoretical Physics
USSR Academy of Sciences

PERGAMON PRESS

OXFORD · NEW YORK · TORONTO · SYDNEY · PARIS · FRANKFURT

U. K.	Pergamon Press Ltd., Headington Hill Hall, Oxford OX3 0BW, England
U. S. A.	Pergamon Press Inc., Maxwell House, Fairview Park, Elmsford, New York 10523, U.S.A.
CANADA	Pergamon Press Canada Ltd., Suite 104, 150 Consumers Road, Willowdale, Ontario M2J 1P9, Canada
AUSTRALIA	Pergamon Press (Aust.) Pty. Ltd., P. O. Box 544, Potts Point, N.S.W. 2011, Australia
FRANCE	Pergamon Press SARL, 24 rue des Ecoles, 75240 Paris, Cedex 05, France
FEDERAL REPUBLIC OF GERMANY	Pergamon Press GmbH, 6242 Kronberg-Taunus, Hammerweg 6, Federal Republic of Germany

First edition 1982

British Library Cataloguing in Publication Data
Sinai, Ya. G.
Theory of phase transitions: Rigorous results
1. Phase transformations (Statistical physics)
I. Title
536′ .401 QC175.16P5 80-40932

ISBN 0-08-026469-7

Co-edition of Akadémiai Kiadó with Pergamon Press Ltd.
Translated by J. Fritz, A. Krámli, P. Major and D. Szász.

Printed in Hungary

CONTENTS

PREFACE

The idea of writing this book arose during my lectures held in the Mathematical Institute of the Hungarian Academy of Sciences, Budapest, on mathematical problems of statistical physics. Their aim was to expound a series of rigorous results about the theory of phase transitions. This book does not give a systematic exposition of the basic theorems of statistical mechanics. However, almost every fact is set forth with such completeness that the reader can grasp the major problems of various topics in the theory of phase transitions. Thus the book is written for those who either are specialists in statistical physics or are inclined to deal with it – directly and seriously.

The book consists of four chapters. In the first, the Hamiltonian, its symmetry group and the limit Gibbs distributions corresponding to a given Hamiltonian are discussed. The discussion is restricted to the lattice case since only this case will be treated later in detail. We give various examples of Hamiltonians: that of the Ising model, those with continuous symmetry, Hamiltonians of some lattice models of quantum field theory, those of the Yang–Mills lattice fields, etc. Further, we treat general results about the existence of limit Gibbs distributions. For example, we show the existence of limit Gibbs distributions for lattice models of quantum field theory.

In the second chapter, phase diagrams of lattice models are considered at low temperatures. The notions of ground state of a Hamiltonian and stability of the set of the ground states of a Hamiltonian are introduced here. In case of periodic configurations, ground states can be defined as configurations with minimal specific energy. The condition of the stability of the ground states, which we call Peierls's condition, requires, roughly speaking, the difference between the energies of a ground state and its local perturbation to be proportional to the boundary separating the domains occupied by different ground states. By assuming the finiteness of the number of ground states and the fulfilment of Peierls's condition, we prove a general statement establishing a connection between the structure of the set of periodic limit Gibbs distributions and that of the set of periodic ground states. This result represents one of the most far-reaching developments of the so-called contour method, proposed by Peierls for proving the existence of long-range order in the Ising model at low temperature. At the end of the chapter, we discuss the notion of ground states for two-dimensional models of quantum field theory. Somewhat surprisingly, we obtain that, asymptotically, when the constant of interaction tends to infinity, the number of ground states does not depend on that part of the Hamiltonian which describes the interaction.

In Chapter III, basic theorems about lattice models with continuous symmetry are given. We prove a theorem of Dobrushin and Shlosman about the absence of spontaneous breakdown of continuous symmetry in two-dimensional models, generalizing a theorem of Mermin–Wagner, and also a theorem of Fröhlich–Simon–Spencer on the presence of spontaneous breakdown of continuous symmetry in models of three or more dimensions with β large. Before the proofs of these theorems, the role of dimensionality is clarified heuristically, in the spirit of Goldstone's general theory.

Chapter IV is devoted to second-order phase transitions and to the theory of scaling probability distributions, connected to these phase transitions. We consider in detail the hierarchical models of Dyson; this example makes it possible to trace the peculiarities of the basic method of the theory: the renormalization group method. The central notion of the theory is that of a scaling probability distribution. These distributions are important because they arise as limit distributions for block spins at the critical point. Scaling distributions can be found easily in the class of stationary Gaussian distributions. Much more difficult is the question about the form of non-Gaussian scaling distributions which arise in the most interesting problems. When constructing such distributions and, in general, in the whole theory, an important role is played by the concept of the linearized renormalization group and its spectrum. For Gaussian scaling distributions, the spectrum of the linearized renormalization group can be computed explicitly. Thanks to this fact, one can find those values of the parameter for which the value 1 appears in the spectrum. In the neighbourhood of these values, we can use the theory of bifurcations and, for non-Gaussian scaling distributions, we can give formal series of the type of the well-known ε-expansions.

At the end of each chapter, bibliographical remarks and some comments are given. The number of papers connected with the theory of phase transitions in statistical mechanics and quantum field theory already reaches some hundreds. Our list of references is intentionally not complete; it contains only directly relevant papers.

The assistance of my colleagues from Budapest, J. Fritz, A. Krámli, P. Major and D. Szász, was of crucial importance in writing this book. Without their enthusiasm and the amount of work they invested in it, this book would not have been written. R. L. Dobrushin read the manuscript and made many useful comments. I also gained much from several discussions with I. M. Lifshitz on the content of the book. From my students, P. M. Bleher, E. I. Dinaburg, E. Gusev, K. Hanin, D. G. Martirosyan, M. Missarov, N. Monina, A. Naimganov and E. Zhalis, I received great help through their remarks on the text and through their participation in producing the manuscript.

CHAPTER I

LIMIT GIBBS DISTRIBUTIONS

1. Hamiltonians

In the theory of random processes, every process is determined by the family of its finite-dimensional probability distributions. By Kolmogorov's fundamental theorem, these distributions give rise to a unique probability measure on the σ-algebra of measurable subsets generated by the finite-dimensional cylindrical sets. The aim of the theory of random processes is to study the properties of a process on the basis of its finite-dimensional distributions. Two typical problems are: what are the properties of typical realizations, and what is the probability of a given behaviour of a random process.

In problems of classical statistical mechanics we find ourselves faced with a different situation. Here the theory is based upon a formal expression, called the Hamiltonian. By its help, all possible conditional probability distributions of the random process or field inside any finite domain can be found under the condition that its values outside the domain are fixed. Now two basic problems appear. The first is to investigate when there exists at least one probability measure with conditional probabilities equal to the expressions given by the help of the Hamiltonian. The second problem is to study the structure of all probability measures having this property.

These problems are natural generalizations of problems of the theory of ordinary Markov chains, where the question about the construction of probability distributions from a system of transition probabilities has also been solved. As we shall see, the theory of finite Markov chains with positive transition probabilities can be embedded into the theory of limit Gibbs distributions as a trivial particular case, and the Hamiltonians can be considered a natural generalization of the transition probabilities or, more exactly, of their logarithms.

Passing on to exact definitions, we shall discuss in detail the case of random fields with discrete time. In general, we shall consider sample spaces Ω consisting of functions $\varphi=\{\varphi(x)\}$ defined on a d-dimensional lattice. In the questions discussed further on, the form of the lattice does not play an important role. Therefore, as usual, we shall consider the case where $x=(x_1, \ldots, x_d)$ varies on the usual integer lattice Z^d with metric $\|x'-x''\|=[\sum_{1\leq i\leq d}(x_i'-x_i'')^2]^{1/2}$ $(x', x''\in Z^d)$. On the other hand, the form of the space where $\varphi(x)$ takes its values may make the theory much easier or much more complicated. In any case, however, we shall always suppose that the range Φ of values of the random variables $\varphi(x)$ does not depend on x, and that it is a measurable space. We list right away the following important particular cases:

1) Φ is a finite set;
2) Φ is a compact metric space; in particular, it is a homogeneous space of a compact Lie group with the natural σ-algebra of Borel subsets;
3) $\Phi = R^1$ or R^n again with the Borel σ-algebra; the latter case is often called a vector model.

Now the sample – or as it is usually called in statistical physics – configuration space Ω itself is a measurable space with a natural σ-algebra γ. The function $\varphi = \{\varphi(x)\}$ is called a *configuration* of the system. The restriction of φ to any subset $V \subset Z^d$ will be denoted by $\varphi(V)$, $\varphi(V) = \{\varphi(x) : x \in V\}$, and $\Omega(V)$ will denote the space of all possible $\varphi(V)$.

Suppose now that, for every non-empty, finite subset $V \subset Z^d$, a function $\mathscr{I}(\varphi(V)) : \Omega(V) \to R^1$ is given. The value of this function is interpreted as the joint interaction energy of the variables $\varphi(x)$ inside the domain V. The family of the functions $\mathscr{I}(\varphi(V))$ is called a potential. For an arbitrary fixed point $x \in Z^d$, we form the sum $\mathscr{U}_x(\varphi) = \sum_{V : x \in V} \frac{1}{|V|} \mathscr{I}(\varphi(V))$, where the summation is taken for all finite domains V containing the point x. It is natural to interpret $\mathscr{U}_x(\varphi)$ as the energy (or potential) of the interaction of the variable $\varphi(x)$ with the variables $\varphi(y)$, $y \in Z^d - \{x\}$.

Generally speaking, the series in the definition of $\mathscr{U}$ can diverge. In case of a finite Φ, however, we shall always consider potentials where this series converges absolutely. For ensuring this convergence, it is sufficient to assume that, for every V,

$$\sup_{\varphi(V)} |\mathscr{I}(\varphi(V))| \leqq \frac{\text{const} \cdot k}{p^{\alpha d} C_{p^d}^{k-2}},$$

where $p = \operatorname{diam} V$, $k = |V|$ and $\alpha > 1$ is a constant.

Pair interaction. If $\mathscr{I}(V) = 0$ unless $|V| = 2$, then F is called a pair or two-body potential. The condition given above requires that for any $V = (x', x'')$

$$|\mathscr{I}(\varphi(x'), \varphi(x''))| \leqq \frac{\text{const}}{\|x' - x''\|^{\alpha d}}, \quad (\alpha > 1).$$

Radius of interaction. Let us suppose that it is possible to find a number R such that $\mathscr{I}(\varphi(V)) \equiv 0$ whenever $\operatorname{diam} V > R$. The minimal number with this property is called the radius of interaction or, briefly, the interaction radius. In this case the series defining $\mathscr{U}$ is finite and makes sense for every φ. If such an R does not exist, we say that the potential has an infinite interaction radius.

If Φ is not compact, and the interaction has an infinite radius, then the series in the definition of $\mathscr{U}$ can also diverge because of the increase of the variables φ at infinity.

For an arbitrary finite subset $W \subset Z^d$, we define the energy $\mathscr{H}(\varphi(W))$ of the configuration φ in the domain W as follows: $\mathscr{H}(\varphi(W)) = \sum_{V \subset W} \mathscr{I}(\varphi(V))$. The sum

$$\mathscr{H}(\varphi(W)|\varphi(Z^d - W)) = \sum_{\substack{V \cap W \neq \emptyset \\ V \cap (Z^d - W) \neq \emptyset}} \mathscr{I}(\varphi(V))$$

will be understood as the interaction energy between the configuration $\varphi(W)$ and the configuration $\varphi(Z^d - W)$, the latter being considered as boundary condition. In case of a compact Φ, the finiteness of this sum follows from our assumption formulated above. If the interaction radius R is finite, then only those subsets can occur in the sum defining $\mathscr{H}(\varphi(W)|\varphi(Z^d - W))$ that are completely contained in an R neighbourhood of W. In the general case, $\mathscr{H}(\varphi(W)|\varphi(Z^d - W))$ is well-defined if the sum is absolutely convergent. Finally, the sum $\mathscr{H}(\varphi(W)) + \mathscr{H}(\varphi(W)|\varphi(Z^d - W))$ will be called total energy of the configuration $\varphi(W)$ under the boundary condition $\varphi(Z^d - W)$.

We shall also consider the formal series

$$\mathscr{H}(\varphi) = \sum_V \mathscr{I}(\varphi(V)) = \sum_{x \in Z^d} \mathscr{U}_x(\varphi)$$

where the first summation goes for all non-empty subsets $V \subset Z^d$. This series is called a Hamiltonian. We can never understand it as a function on the space Ω, but, fortunately, we do not need this either. What is of importance for us is that $\mathscr{I}$ defines $\mathscr{H}(\varphi(W))$ and $\mathscr{H}(\varphi(W)|\varphi(Z^d - W))$ uniquely. Often we can also consider differences $\mathscr{H}(\varphi') - \mathscr{H}(\varphi'')$ for configurations φ', φ'' which are equal almost everywhere, i.e. everywhere except for a finite number of points.

Symmetry groups of a Hamiltonian. A quite important notion for many examples and applications of the theory is the symmetry group of a Hamiltonian. We begin with some examples.

Let $\{T^y, y \in Z^d\}$ be the group of spatial translations in the space Ω, i.e. $(T^y \varphi)(x) = \varphi(x-y)$. A Hamiltonian is called translation invariant if $\mathscr{H}(\varphi) = \mathscr{H}(T^y \varphi)$ for any $y \in Z^d$, i.e. if $\mathscr{I}(\varphi(V)) = \mathscr{I}((T^y \varphi)(V+y))$. In other words, the function $\mathscr{I}(\varphi(V))$ takes on the same values for congruent subsets V. For a pair interaction, translation invariance means that $\mathscr{I}(\varphi(x'), \varphi(x''))$ depends on the difference $x' - x''$ only.

More generally, let Z_0^d be a subgroup of Z^d and let $\{T^y, y \in Z_0^d\}$ be the corresponding subgroup of the group of lattice translations. Suppose that the index of the subgroup is finite, i.e. the quotient group Z^d/Z_0^d is finite. The Hamiltonian $\mathscr{H}$ is called periodic (or more exactly Z_0^d-periodic) if $\mathscr{H}(\varphi) = \mathscr{H}(T^y \varphi)$ for every $y \in Z_0^d$. This means that, for any fixed V, the function $\mathscr{I}(\varphi(V+y))$ depends on the coset determined by y with respect to the subgroup Z_0^d.

Suppose now that a group G acts in the range space Φ. Extend its action to the whole Ω by putting $(g\varphi)(x)=g\varphi(x)$ for every $g\in G$. We say that the Hamiltonian $\mathscr{H}(\varphi)$ is G-invariant if, for any $g\in G$, $\mathscr{H}(g\varphi)=\mathscr{H}(\varphi)$, i.e.

$$\mathscr{I}(\varphi(V)) = \mathscr{I}((g\varphi)(V)).$$

(Basic) Example 1. $\Phi=\{-1, 1\}$. The group G consists of two elements: the identity transformation e and the symmetry $g\varphi=-\varphi$, that is $(g\varphi)(x)=-\varphi(x)$ for any $x\in Z^d$. This symmetry is called $\pm$ symmetry (or invariance).

If $\Phi=-\Phi\subset R^1$, then the Hamiltonian of any pair interaction of the form

$$\mathscr{H}(\varphi) = \sum \mathscr{I}(x', x'')\varphi(x')\varphi(x'')$$

is G-invariant.

Example 2. Let $\Phi=S^1$ (S^{n-1} denotes the unit sphere of R^n), and let G be the commutative rotation group of the unit circle. G will also denote the action of G on Ω. Then it is obvious that the Hamiltonian $\mathscr{H}(\varphi)=\sum\mathscr{I}(x', x'')(\varphi(x'), \varphi(x''))$ is G-invariant ((,) is the scalar product).

We introduce a more general definition. Let G be a topological group acting measurably in the space Φ. This means that, for any measurable function f over Φ, the function $f(g\varphi)$ is measurable on the product space $G\times\Phi$.

Let us consider the group $\hat{G}$ of transformations of the space Ω; a transformation $\hat{g}\in\hat{G}$ is given by a point $z\in Z^d$ and a G-valued function $g(y)$ defined on Z^d. The transformation $\hat{g}=(z, g(y))$ acts on the configuration φ as $(\hat{g}\varphi)(x)=g(x)\varphi(x-z)$, $x\in Z^d$. Two transformations can be composed by the following law

$$(z, g(y))\circ(z', g'(y)) = ((z+z', g(y)g'(y-z)).$$

$\hat{G}$ has a natural topology making it a topological group. In this topology, the sequence $\hat{g}_n=(z_n, g_n(y))\in\hat{G}$ converges to an element $\hat{g}=(z, g(y))$ if there exists an integer N such that $z_n=z$ whenever $n>N$, and, moreover, if $g_n(y)$ converges to $g(y)$ uniformly in $y\in Z^d$, as $n\to\infty$.

For an arbitrary finite subset $V\in Z^d$ and a transformation $\hat{g}=(z, g)\in\hat{G}$, we have the obvious relation

$$(\hat{g}\varphi)(V) = ((\hat{g}\varphi)(y), y\in V) = (g(y)\varphi(y-z), y\in V) =$$
$$= (g(z+y)\varphi(y), y\in V-z).$$

Let us denote this latter expression by $\hat{g}\varphi(V-z)$.

Definition 1.1. *Let S be a closed subgroup of $\hat{G}$. We say the potential $\mathscr{H}(\varphi)$ is S-invariant if, for any $\hat{g}=(z, g)\in S$, for arbitrary finite subset V and any configuration $\varphi\in\Omega$, we have*

$$\mathscr{I}(\varphi(V)) = \mathscr{I}((\hat{g}\varphi)(V-z)).$$

Now we are going to give examples of Hamiltonians with various symmetry groups.

2. Examples of Hamiltonians

1. One-dimensional Markov chains. Let Ω be the sample space of a finite Markov chain with r states. That is, the range space Φ consists of r elements and P is a measure on Ω corresponding to a homogeneous Markov chain with a stationary transition matrix $\Pi = \|\pi_{ij}\|$ and with stationary distribution $\pi = \{\pi_1, \ldots, \pi_r\}$. We restrict ourselves to the case when all $\pi_{ij} > 0$. For $V = (k, k+1, \ldots, k+m)$, the probability of an arbitrary configuration $\varphi(V) = (\varphi(k), \varphi(k+1), \ldots, \varphi(k+m))$ is equal to

$$\pi_{\varphi(k)} \cdot \pi_{\varphi(k)\varphi(k+1)} \cdot \pi_{\varphi(k+1)\varphi(k+2)} \cdots \pi_{\varphi(k+m-1)\varphi(k+m)} =$$
$$= \exp\{\ln \pi_{\varphi(k)} + \sum_{i=k}^{k+m-1} \ln \pi_{\varphi(i)\varphi(i+1)}\}.$$

Introducing the Hamiltonian $\mathscr{H}(\varphi) = -\sum_{i=-\infty}^{\infty} \ln \pi_{\varphi(i)\varphi(i+1)}$, we see that the interaction energy $\mathscr{I}(\varphi(V))$ differs from zero only if $V = (i, i+1)$ and then $\mathscr{I}(\varphi(i), \varphi(i+1)) = -\ln \pi_{\varphi(i)\varphi(i+1)}$.

2. *d*-dimensional Ising model. Let the range space Φ consist of two points: $\Phi = \{1, -1\}$. Consider the Hamiltonian $\mathscr{H}$ for which $\mathscr{I}(\varphi(V)) = 0$ unless $V = \{x, y\}$, $\|x-y\| = 1$, and, in this case, let $\mathscr{I}(\varphi(V)) = \mathscr{I}\varphi(x)\varphi(y)$, i.e.

$$\mathscr{H} = -\mathscr{I} \sum_{\{x,y\}:\|x-y\|=1} \varphi(x)\varphi(y).$$

A model given by this potential is called a *d*-dimensional Ising model (with zero external field). If $\mathscr{I} > 0$, the model is called ferromagnetic, while if $\mathscr{I} < 0$, it is called antiferromagnetic. The reason for these attributes will be given later. It is clear that $\mathscr{H}$ is translation-invariant and, moreover, that it has Z_2 as a symmetry group. Here Z_2 consists of two elements: *Id*, the identity transformation, and *s*, the symmetry transformation defined by $(s\varphi)(x) = -\varphi(x)$. This group is nothing other than the $\pm$-symmetry group mentioned above.

The model given by the Hamiltonian

$$\mathscr{H} = -\mathscr{I} \sum_{\|x-y\|=1} \varphi(x)\varphi(y) - h \sum_{x \in Z^d} \varphi(x)$$

is called an Ising model with external field h. In this case, the Hamiltonian is only translation-invariant.

3. *X–Y* model. Let $d=2$, and let the range space Φ be the surface S^1 of the unit circle. It is convenient to conceive a configuration of the model as a family of unit vectors starting out at each point x of the lattice Z^2 and to understand $\varphi(x)$ as the angle between the vector starting out at x and the positive horizontal axis of the coordinate system $(-\pi < \varphi(x) \leqq \pi)$. Take some even function $\mathscr{U}(\varphi)$ defined on S^1, and consider the Hamiltonian $\mathscr{H} = \sum_{\{x,y\}:\|x-y\|=1} \mathscr{U}(\varphi(x) - \varphi(y))$. We obtain the *X–Y* model if we put, in particular, $\mathscr{U}(\varphi(x) - \varphi(y)) = -\cos(\varphi(x) - \varphi(y))$. The model where $\mathscr{U}(\varphi)$ is proportional to the logarithm of the θ-function is also of great interest. We recall the definition of the θ-function

$$\theta(\varphi) = \sum_{n=-\infty}^{\infty} e^{-n^2} e^{in\varphi} = C \sum_{n=-\infty}^{\infty} e^{-(\varphi-2\pi n)^2}.$$

The Hamiltonian $\mathscr{H}$ is, obviously, translation-invariant. Moreover, it is also invariant with respect to the group S^1 where, for every $g \in S^1$, we have $(g\varphi)(x) = \varphi(x)+g$. Here g is a number, $0 \leqq g < 2\pi$, and the addition is understood mod 2π. This is the simplest example for a model with continuous symmetry.

4. Classical rotator. Let $d \geqq 1$ be arbitrary, and let the range space Φ be the surface of the m-dimensional sphere: $\Phi = S^{m-1}$. The model with the Hamiltonian $\mathscr{H} = -\mathscr{I} \sum_{\|x-y\|=1} (\varphi(x), \varphi(y))$ is called a classical rotator. This Hamiltonian is invariant with respect to the rotation group $O(m)$, where, for any $g \in O(m)$, we put $(g\varphi)(x) = g\varphi(x)$.

5. Yang–Mills lattice field. Instead of Z^d, we shall now consider for any $d \geqq 2$ the set whose elements are the bonds of the lattice Z^d. Any of these bonds can be described by an ordered pair $l=(x, y)$, $\|x-y\|=1$, where x is the starting point, and y is the endpoint of the bond. The ordered pair (y, x) will be understood as the point $-l$. The range space Φ will now be the group $SO(n)$. Consider the space of configurations $\varphi = \{\varphi(l)\}$ which satisfy the condition $\varphi(-l) = \varphi^{-1}(l)$ for every l. The following Hamiltonian is called a Yang–Mills Hamiltonian:

$$\mathscr{H} = -\sum \operatorname{tr}(\varphi(x_1, x_2) \cdot \varphi(x_2, x_3) \cdot \varphi(x_3, x_4) \cdot \varphi(x_4, x_1)),$$

where (x_1, x_2, x_3, x_4) is a 4-tuple of vertices such that they determine a two-dimensional face of a Z^d lattice cell, and $x_1 \to x_2 \to x_3 \to x_4 \to x_1$ is the positive orientation of the face in question; the summation goes over all such faces. In the definition of the Hamiltonian, tr can be substituted by an arbitrary character of the group $SO(n)$.

The Yang–Mills Hamiltonian possesses a very large symmetry group. In fact, let $g = \{g(x)\}$ be an arbitrary function from Z^d into the group $SO(n)$. For $l=(x, y)$, put

$$(g\varphi)(l) = g^{-1}(x)\varphi(l)g(y).$$

Then each summand in the definition of $\mathscr{H}$ and hence the whole Hamiltonian $\mathscr{H}$ remain invariant under the action of g. Consequently, $\mathscr{H}$ is invariant with respect to the group $\prod_{x \in Z^d} SO(n)(x)$.

It is natural to conceive the Yang–Mills field as follows: let us assume that a copy $R^n(x)$ of the n-dimensional Euclidean space is attached to every point x of the lattice Z^d. $\varphi(l)$ with $l=(x, y)$ will be understood as an isometry between $R^n(x)$ and $R^n(y)$ and the whole configuration $\varphi = \{\varphi(l)\}$ as a connection in the whole set of these spaces. The Hamiltonian remains invariant if any $R^n(x)$ is transformed by the automorphism generated by $g(x)$.

3. The definition of limit Gibbs distributions

Suppose we are given a Hamiltonian $\mathscr{H}$ and a (not necessarily normed) measure χ defined on Φ. For every V, we can take the product measure $\Pi_{s\in V} d\chi(\varphi(s))$. For an arbitrary finite V, we shall consider configurations $\varphi(Z^d-V)$ such that $\mathscr{H}(\varphi(V)|\varphi(Z^d-V))$ is finite for every configuration $\varphi(V)$ and the integral

$$\Theta = \int e^{-\mathscr{H}_V(\varphi)} \Pi_{s\in V} d\chi(\varphi(s)) \tag{1.1}$$

is also finite, where $\mathscr{H}_V(\varphi)=\mathscr{H}(\varphi(V))+\mathscr{H}(\varphi(V)|\varphi(Z^d-V))$. This integral is called the partition function or statistical sum. (We remark that the partition function is usually denoted by Z or Ξ or Θ. We shall also use the notation Z to emphasize the difference between partition functions of different models.)

Definition 1.2. *The conditional Gibbs distribution for the domain V under the boundary condition $\varphi(Z^d-V)$ is a probability distribution on the space $\Omega(V)$ of configurations $\varphi(V)$ whose density with respect to $\Pi_{s\in V} d\chi(\varphi(s))$ is of the form*

$$p(\varphi(V)|\varphi(Z^d-V)) = \frac{e^{-\mathscr{H}_V(\varphi)}}{\Theta}. \tag{1.2}$$

In case of a compact range space Φ, $\mathscr{H}_V(\varphi)$ makes sense for every $\varphi\in\Omega$ if $\mathscr{U}_x(\varphi)$ is continuous.

Let $V_1\subset V_2$ and let the configurations $\varphi(V_1)$, $\varphi(V_2-V_1)$, $\varphi(Z^d-V_2)$ be given. Then a straightforward consequence of definition 1.2 is that

$$p(\varphi(V_1)|\varphi(Z^d-V_1)) = \frac{p(\varphi(V_2)|\varphi(Z^d-V_2))}{p(\varphi(V_2-V_1)|\varphi(Z^d-V_2))}.$$

The integral in the denominator is obviously equal to

$$\int p(V_2)|\varphi(Z^d-V_2))\, \Pi_{s\in V_1} d\chi(\varphi(s)).$$

We remark that the same equation should be satisfied (almost everywhere) for the conditional distributions $p(\varphi(V)|\varphi(Z^d-V))$ corresponding to an arbitrary probability distribution P given on Ω.

The following definition is central to the whole theory.

Definition 1.3. *The probability distribution P given on the space Ω is a limit Gibbs distribution corresponding to the Hamiltonian $\mathscr{H}$ if, for any finite $V\subset Z^d$,*

1) *$\mathscr{H}(\varphi(V)|\varphi(Z^d-V)$ and Θ (see (1.1)) are finite with P-probability 1 (i.e. for P-almost every $\varphi\in\Omega$);*

2) *with P-probability 1, the conditional distribution induced by P on $\Omega(V)$ under a fixed boundary condition $\varphi(Z^d-V)$ is absolutely continuous with respect to the measure*

$\Pi_{s\in V} d\chi(\varphi(s))$, and its density with respect to the same measure equals

$$p(\varphi(V)|\varphi(Z^d-V)) = \Theta^{-1}\exp\{-\mathscr{H}_V(\varphi)\} = \tag{1.3}$$
$$= \Theta^{-1}\exp\{-\mathscr{H}(\varphi(V))-\mathscr{H}(\varphi(V)|\varphi(Z^d-V))\}.$$

In other words, this conditional distribution should be a conditional Gibbs distribution under the boundary condition $\varphi(Z^d-V)$ with P-probability 1.

If Φ is compact, and all the functions $\mathscr{I}(\varphi(V))$ are continuous, then (1.1), (1.2) and (1.3) have sense for every configuration $\varphi(Z^d-V)$. At the same time the general theory of probability ensures the existence of the conditional probabilities almost everywhere only. Consequently, our definition 1.3 requires that an expression defined almost everywhere be equal to an expression defined everywhere. This can be explained by the fact that measure theory is always constructed "up to sets of measure 0".

The fundamental problem of statistical physics is to describe all limit Gibbs measures for a given Hamiltonian. This problem can only be solved in particular, relatively simple cases. In a sense, this whole monograph is devoted to the discussion of various known rigorous results connected with this problem.

The definitions 1.2, 1.3 can be extended in a natural way. Let μ_0 be an arbitrary probability measure on Ω. For any finite set V, we introduce the conditional distributions $\mu_0(\cdot|\varphi(Z^d-V))$ given on the configuration space $\Omega(V)$. Again, these distributions are defined only almost everywhere. However, we assume they can be defined everywhere, and satisfy the relation

$$\int_{C_2}\mu_0(C_1|\varphi(Z^d-V_1))\,d\mu_0(\varphi(V_2-V_1)|\varphi(Z^d-V_2)) =$$
$$= \mu_0(C_1\times C_2|\varphi(Z^d-V_2))$$

for arbitrary finite $V_1\subset V_2$. Here C_1, C_2 are arbitrary measurable subsets in the spaces $\Omega(V_1)$ and $\Omega(V_2-V_1)$ respectively, and $d\mu_0(\varphi(V_2-V_1)|\varphi(Z^d-V_2))$ is the conditional probability distribution on the configurations $\varphi(V_2-V_1)$ under the fixed configuration $\varphi(Z^d-V_2)$.

Instead of (1.1) and (1.2), consider now

$$\theta = \Theta(\varphi(Z^d-V)) = \tag{1.1'}$$
$$= \int \exp\{-\mathscr{H}(\varphi(V)|\varphi(Z^d-V))-\mathscr{H}(\varphi(V))\}\,d\mu_0(\varphi(V)|\varphi(Z^d-V))$$

$$p(\varphi(V)|\varphi(Z^d-V)) = \Theta^{-1}\exp\{-\mathscr{H}(\varphi(V)|\varphi(Z^d-V))-\mathscr{H}(\varphi(V))\} \tag{1.2'}$$

and consider conditional distributions on the configuration spaces $\Omega(V)$, absolutely continuous with respect to $\mu_0(\cdot|\varphi(Z^d-V))$ with densities (with respect to $\mu_0(\cdot|\varphi(Z^d-V))$) equal to (1.2'). Then it is natural to give the following

Definition 1.3′. *We say that P is a limit Gibbs distribution corresponding to the Hamiltonian* $\mathscr{H}$ *and the measure* μ_0 *if its conditional probability distributions are absolutely continuous with respect to* $\mu_0(\cdot|\varphi(Z^d-V))$ *with densities equal to* (1.2′) *almost everywhere.*

For μ_0 we can take, for example, any limit Gibbs distribution corresponding to a potential with a finite interaction radius. In principle, the definitions 1.3 and 1.3′ show a natural way of transition from one probability measure to an other one.

Limit Gibbs distributions and symmetry groups of the Hamiltonians. Let S be a symmetry group of the Hamiltonian $\mathscr{H}$. Suppose the measure μ_0 is also invariant with respect to the group S. Then under some natural additional assumptions, the group S acts on the set of limit Gibbs distributions, too. Consequently, the whole set of limit Gibbs distributions can be decomposed into orbits of the group S.

Periodic boundary conditions. In many cases, it is necessary to consider Gibbs distributions with the so-called periodic boundary conditions. We show how these distributions can be constructed. Take a d-dimensional parallelepiped with one vertex in the origin $0=(0, \ldots, 0)$ of the lattice. The vertices of V generate a subgroup $Z_0^d \subset Z^d$ of finite index. Choose a periodic configuration $\varphi=\{\varphi(x), x\in Z^d\}$ with period V. For such configurations, $T^t\varphi=\varphi$ for any $t\in Z_0^d$. In case of a V-periodic configuration, the potential $\mathscr{U}_x(\varphi)$ is a V-periodic function on the lattice. We call a probability distribution, given on the space of V-periodic configurations, a Gibbs distribution in the domain V with periodic boundary conditions if its density with respect to the measure $\prod_{x\in V} d\chi(\varphi(s))$ is equal to $\Theta^{-1}\exp\{-\sum_{x\in V}\mathscr{U}_x(\varphi)\}$, where Θ is the corresponding norming constant, i.e. $\Theta=\int \exp\{-\sum_{x\in V}\mathscr{U}_x(\varphi)\}\prod_{s\in V} d\chi(\varphi(s))$. It is assumed that $\Theta<\infty$.

Free boundary conditions. We often consider Gibbs distributions in the domain V with the so-called free boundary conditions. It is the probability distribution on the space $\Omega(V)$ whose density with respect to the measure $\prod_{s\in V} d\chi(\varphi(s))$ is of the form $\Theta^{-1}\exp\{-\mathscr{H}(\varphi(V))\}$.

Limit Gibbs distributions are constructed by Hamiltonians or by potentials defining these Hamiltonians. Thus, it is natural to consider the Hamiltonian as a parameter determining the limit Gibbs distributions. These parameters are sufficiently simple, and the task of the theory of limit Gibbs distributions can be formulated so as to study the properties of the constructed probability distributions as functions of the Hamiltonians.

4. Examples

Markov chains as limit Gibbs distributions

The Hamiltonian corresponding to a Markov chain has been described in Section 2. For the measure χ, let us choose the Bernoulli measure for which each value of $\varphi(x)$ has probability $1/r$ (r is the number of states). Then the stationary Markov chain is a limit Gibbs distribution. The usual ergodic theorem for Markov chains can be formulated in such a way that it implies, in this case, the uniqueness of the limit Gibbs distribution.

Systems with finite interaction radius as Markov fields with multidimensional time

If the Hamiltonian $\mathscr{H}$ has a finite interaction radius, and Φ is a finite set, then the conditional probabilities (1.2) are defined everywhere. Expression (1.2) shows that the conditional probability of a configuration $\varphi(V)$ does not depend on the whole configuration $\varphi(Z^d - V)$ but only on its restriction to the R-neighbourhood of the boundary of V. Here, as usual, R stands for the interaction radius. Thus, limit Gibbs distributions for such Hamiltonians can be considered as Markov fields of memory of length R with multidimensional time. The theory of such fields for $d>1$ substantially differs from that for $d=1$, i.e. from the theory of Markov chains of order R. In the next chapter, it will be shown that for $d>1$ there are many natural cases when several limit Gibbs distributions appear corresponding to a given Hamiltonian.

Stationary Gaussian fields as limit Gibbs distributions

Consider a stationary Gaussian field on the d-dimensional lattice Z^d. From our point of view, the sample space Ω of such a field consists of infinite configurations $\varphi = \{\varphi(x), x \in Z^d\}$ where any single variable $\varphi(x)$ can take on arbitrary real values. By definition, a probability distribution G on Ω is called Gaussian if any r-tuple of variables $(\varphi(x_1), \ldots, \varphi(x_r))$ obeys an r-dimensional Gaussian distribution. We suppose that $E\varphi(x)=0$, $E\varphi(x)\varphi(y)=b(x-y)=\int_{[0,1]^d} \exp\{2\pi i(\lambda, x-y)\}\, \varrho(\lambda)\, d\lambda$. The function $\varrho(\lambda)$ is the spectral density of the Gaussian distribution G. It is non-negative and $\int \varrho(\lambda)\, d\lambda < \infty$.

Suppose that $1/\varrho(\lambda)$ is a continuous function on $[0, 1]^d$ with absolutely convergent Fourier series, i.e. $\dfrac{1}{\varrho(\lambda)} = \sum_{x \in Z^d} a(x) \exp\{-2\pi i(\lambda, x)\}$, and

$$A = \sum_{x \in Z^d} |a(x)| < \infty.$$

Theorem 1.1. *The Gaussian distribution G is a limit Gibbs distribution corresponding to the potential*

(1.4)
$$\mathcal{H} = \frac{1}{2} \sum_{x,y\in Z^d} a(x-y)\,\varphi(x)\,\varphi(y)$$

provided the role of the measure χ is played by the Lebesgue measure on the line.

Proof. Let $V \subset Z^d$ be a fixed, finite subset and, for every $t\in Z^d$, consider the series $\psi(t) = \sum_{y\in Z^d-V} a(t-y)\varphi(y)$. This series converges in $\mathcal{L}^2(\Omega, \gamma, G)$, because the formal expression for $E\psi^2(t) = \sum_{y_1,y_2\in Z^d-V} a(t-y_1)a(t-y_2)b(y_2-y_1)$ is finite. This can be seen from the following inequalities

$$\sum_{y_1,y_2\in Z^d} |a(t-y_1)a(t-y_2)b(y_2-y_1)| \leqq$$
$$\leqq b(0) \sum_{y_1,y_2\in Z^d} |a(t-y_1)|\,|a(t-y_2)| \leqq b(0)\,A^2.$$

Consequently, each $\varphi(t)$ can be considered as a Gaussian random variable, finite with probability 1. We remark that, in the Gaussian case, the condition $\sum_{x\in Z^d} |a(x)| < \infty$ also ensures the almost certain convergence of the series defining $\psi(t)$.

For a limit Gibbs distribution corresponding to the Hamiltonian (1.4), the conditional density of $\varphi(V)$ under fixed $\varphi(x)$, $x\in Z^d - V$ has the form

(1.5)
$$(\Theta(\psi))^{-1} \exp\left\{-\frac{1}{2}\left[\sum_{x,y\in V} a(x-y)\varphi(x)\varphi(y) + 2\sum_{x\in V} \varphi(x)\psi(x)\right]\right\} =$$
$$= \text{const}\cdot\exp\left\{-\frac{1}{2}(B(\varphi + B^{-1}\psi),\ \varphi + B^{-1}\psi)\right\}.$$

In other words, it is a multidimensional Gaussian density with expectation vector $-B^{-1}\psi$ and covariance matrix $B = \|a(x-y)\|_{x,y\in V}$.

We assert that, for the Gaussian distribution G, the conditional density of the variables $\varphi(x)$, $x\in V$ under fixed $\varphi(y)$, $y\in Z^d - V$ is equal to (1.5). Let us show first that the random variables $\varphi(x) + (B^{-1}\psi)(x)$, $x\in V$ are independent of the variables $\varphi(y)$, $y\in Z^d - V$ (with respect to the distribution G). Since B is not singular, it is sufficient to check the independence of

$$(B\varphi)(x) + \psi(x) = \sum_{t\in V} a(x-t)\varphi(t) + \sum_{t\in Z^d-V} a(x-t)\varphi(t) =$$
$$= \sum_{t\in Z^d} a(x-t)\varphi(t).$$

For every $y\in Z^d - V$, we have

$$E\left(\sum a(x-t)\varphi(t),\ \varphi(y)\right) = \sum a(x-t)\,b(t-y).$$

This last series converges absolutely, and hence it is equal to δ_{xy}. Since $x\in V$, $y\in Z^d - V$, it is equal to 0.

The fact we proved above implies that the vector $-B^{-1}\psi$ is the conditional expectation vector of the variables $\varphi(x)$, $x\in V$ under fixed $\varphi(y)$, $y\in Z^d-V$.

It remains to prove that the covariance matrix of the vector $\varphi'=\{\varphi(x)+(B^{-1}\psi)(x), x\in V\}$ is B^{-1}. Let

$$\varphi'' = B\varphi' = \{\textstyle\sum_{y\in V} a(x-y)\varphi(y)+\psi(x), x\in V\} = $$
$$= \{\textstyle\sum_{y\in Z^d} a(x-y)\varphi(y), x\in V\}.$$

The fact to be proved is equivalent to the statement that the covariance matrix of the vector φ'' is B.

However,

$$E\varphi''(x_1)\varphi''(x_2) = \textstyle\sum_{y_1,y_2\in Z^d} a(x_1-y_1)b(y_1-y_2)a(y_2-x_2) =$$
$$= \textstyle\sum_{y_1\in Z^d} a(x_1-y_1)\delta_{y_1x_2} = a(x_1-x_2).$$

The series in the last relation converges absolutely ensuring the legality of our transformations. Thus the covariance matrix of the vector φ'' is B, and so that of φ' is B^{-1}. Q.e.d.

This example is instructive in many aspects. First of all, (1.5) does not make sense for all boundary conditions $\varphi(Z^d-V)$ but only for those where every $\psi(x)$, $x\in V$, is finite. Secondly, quadratic Hamiltonians arise in many problems in quantum field theory and in statistical physics (Hamiltonians of free fields), and the variables $\varphi(x)$ take on real values. The simplest example of such a Hamiltonian is $\mathscr{H}= =\int_{R^d}((\nabla\varphi)^2+m_0^2\varphi^2)dx$. Its lattice analogue for a lattice constant h has the form

$$\mathscr{H} = \textstyle\sum_{x\in hZ^d}\left[\sum_{i=1}^d h^{-2}(\varphi(x_1, \dots, x_i+h, \dots, x_d)-\varphi(x_1, \dots, x_d))^2+\right.$$
$$\left.+m_0^2\varphi(x)\right]h^d.$$

The outer sum is taken for points $x=(x_1, \dots, x_d)$ of the form $x_i=n_ih$, n_i integer, $-\infty<n_i<\infty$. In the lattice case, as we have shown, the stationary Gaussian distributions belong to the family of limit Gibbs fields corresponding to these Hamiltonians.

Lattice models for two-dimensional quantum field theory

Let $d=2$. Consider, as at the end of the previous example, the lattice hZ^2 with lattice constant h. Here we shall use the construction of limit Gibbs distributions described in Definition 1.3′. Namely, the role of the measure μ_0 will be played by a Gaussian distribution with mean 0 corresponding to the Hamiltonian

$$\mathscr{H}_0 = \frac{1}{2}\textstyle\sum_x [(\varphi(x_1+h, x_2)-\varphi(x_1, x_2))^2+(\varphi(x_1, x_2+h)-\varphi(x_1, x_2))^2+$$
$$+m_0^2h^2\varphi^2(x_1, x_2)],$$

where x runs through hZ^2. The Hamiltonian $\mathcal{H}_0$ has a finite interaction radius. Thus, the corresponding conditional distributions $\mu_0(\cdot|\varphi(Z^d-V))$ are defined everywhere and depend only on a finite number of points. Consequently, using μ_0, we can construct limit Gibbs distributions (cf. Definition 1.3′).

The random variable $\varphi(x)$ has a Gaussian distribution with mean 0 and variance $\sigma_h \sim \text{const} \cdot \ln h^{-1}$ (for h small). For any integer $p > 1$, denote by $G_p(t) = t^p + \dots$ the usual Hermite polynomial of order p with leading coefficient equal to 1. Denote by $G_p^{(h)}(t)$ the Hermite polynomial $\sigma_h^{p/2} G_p(t/\sqrt{\sigma_h}) = t^p + \dots$, and introduce the Hamiltonian $\mathcal{H}^{(\text{int})} = \sum_{p=1}^{2r} a_p \sum_{x \in hZ^2} G_p^{(h)}(\varphi(x))$, where a_p are constants, $a_{2r} > 0$. In quantum field theory, Wick notation is generally used, in which $G_p^{(h)}(\varphi(x)) = :\varphi^p(x):_{\sigma_h}$. The notation: $:_{\sigma_h}$ means that the coefficients of the Hermite polynomials are constructed according to a Gaussian distribution of variance σ_h. (Properties of Hermite polynomials will be discussed in more detail in Chapter IV.) Limit Gibbs distributions arise in a natural way when we study models of two-dimensional quantum field theory. They are constructed by the aid of the free field μ_0 and the interaction Hamiltonian $\mathcal{H}^{(\text{int})}$. If $a_p = 0$ for odd p, then the Hamiltonian possesses the $\pm$ symmetry. Another possibility for a more direct construction of the limit Gibbs distribution would be to take the direct product of Lebesgue measures for μ_0, and to take

$$\mathcal{H} = \mathcal{H}_0 + \mathcal{H}^{(\text{int})}$$

for the Hamiltonian. The formal passing to a limit as $h \to 0$ leads to the Hamiltonian of a continuous field:

$$\mathcal{H} = \iint [(\nabla\varphi, \nabla\varphi) + m_0^2 \varphi^2 + \lambda : P(\varphi) :] \, dx_1 \, dx_2$$

with $P(\varphi) = \sum_{p=1}^{2r} a_p \varphi^p$. The coefficient m_0 is called the bare mass or the mass of the free field, and λ the interaction constant. The properties of limit Gibbs distributions depend on the dimensionless parameter λ/m_0^2.

Berezinsky's transformation of the X–Y model

Let V be a rectangle on the plane. Consider the X–Y model with potential $\mathcal{U}$, and take the Gibbs distribution in the domain V with free boundary conditions. For reference measure χ_V, we choose the direct product of Haar measures. Then the partition function is of the form

$$\Theta = \int \prod_{x \in V} d\varphi(x) \prod_{\|x'-x''\|=1} \exp\{-\mathcal{U}(\varphi(x') - \varphi(x''))\}. \tag{1.6}$$

Now the main point is to find a convenient transcription of the latter product. This product is taken over unordered pairs $\{x', x''\} : \|x' - x''\| = 1$, determining the edges of the lattice. Suppose every such edge has a fixed orientation. The oriented edge will be denoted by l, i.e. $l = (x', x'')$, and we call x' the starting point and x'' the endpoint of l. The oriented edge (x'', x') will be denoted by $-l$.

The product $\Pi_{\|x'-x''\|=1} \exp\{-\mathscr{U}(\varphi(x')-\varphi(x''))\}$ can now be considered as follows: we choose a family $\mathscr{L}$ of oriented edges $l=(x', x'')$ in such a way that of each pair $l, -l$ exactly one edge belongs to $\mathscr{L}$; the product is taken for edges in $\mathscr{L}$, and in the factor $\exp\{-\mathscr{U}(\varphi(x')-\varphi(x''))\}$ we write the starting point first and the endpoint second.

Suppose $\mathscr{U}$ is continuous on the (boundary of the) unit circle. Then $\exp\{-\mathscr{U}\}$ is also continuous and it can be expanded in a Fourier series

$$\exp\{-\mathscr{U}(\varphi)\} = \sum_{n=-\infty}^{\infty} c_n e^{2\pi i n\varphi}.$$

Put this expansion into (1.6) and interchange the order of summation and product as follows:

$$\begin{aligned} \Theta &= \int \Pi_{x\in V} d\varphi(x)\, \Pi_{\|x'-x''\|=1} \exp\{-\mathscr{U}(\varphi(x')-\varphi(x''))\} = \\ &= \int \Pi_{x\in V} d\varphi(x)\, \Pi_{l\in\mathscr{L}} \sum_n c_n \exp\{2\pi i n(\varphi(x')-\varphi(x''))\} = \\ &= \sum_n \Pi_{l\in\mathscr{L}} c_{n(l)} \int \Pi_{x\in V} d\varphi(x) \exp\{2\pi i\varphi(x)\, \Delta n(x)\}. \end{aligned} \tag{1.7}$$

To make the last step clear it is natural to understand the index n in the sum $\sum_n c_n \exp\{2\pi i n(\varphi(x')-\varphi(x''))\}$ as one depending on l and to understand the summation $\sum_n$ in the last row as one going over all possible integer-valued functions $n(l)$ defined on $\mathscr{L}$. The factor $\Delta n(x)$ is a sum for all edges $l\in\mathscr{L}$ containing x as a vertex and, moreover, such a summand is taken with $+$ sign or $-$ sign according to whether x is the starting point or the endpoint of the edge.

Every function $n(l)$ defined on $\mathscr{L}$ can be uniquely extended to a function defined on the set $\mathscr{L}^{(o)}$ of all oriented edges of the lattice in the rectangle V by setting $n(l)= -n(-l)$ for every $l\notin\mathscr{L}$. Then $\Delta n(x)$ can be written in an especially simple form

$$\Delta n(x) = \sum n(l), \tag{1.8}$$

where the summation goes over the four edges $l\in\mathscr{L}^{(o)}$ containing x as starting point.

Compute the integrals in (1.7). Then, if $\Delta n(x)\neq 0$ for at least one site x, the integral vanishes. Otherwise it equals 1. This results in

$$\Theta = \sum_{\substack{\Delta n(x)=0 \\ \text{for every } x\in V}} \Pi_{l\in\mathscr{L}} c_{n(l)}. \tag{1.9}$$

In other words, the summation in (1.9) should be taken for all functions $n(l)$ defined on the set of oriented edges $l\in\mathscr{L}^{(o)}$ and satisfying the conditions $n(l)=-n(-l)$, for every $l\in\mathscr{L}^{(o)}$, $\Delta n(x)=0$ for every $x\in V$, where $\Delta n(x)$ is as in (1.8). Moreover, each summand of this sum has the weight $\Pi_{l\in\mathscr{L}} c_{n(l)}$. $\mathscr{U}$ and thus $e^{-\mathscr{U}}$ are even, implying $c_{n(-l)}=c_{n(l)}$. Consequently

$$\Pi_{l\in\mathscr{L}} c_{n(l)} = (\Pi_{l\in\mathscr{L}^{(o)}} c_{n(l)})^{1/2}.$$

If every $c_{n(l)} > 0$, as, for example, in the case when $\mathscr{U}$ is the logarithm of the θ-function, then

$$\left(\Pi_{l\in\mathscr{L}^{(o)}} c_{n(l)}\right)^{1/2} = \exp\left\{\frac{1}{2}\sum_{l\in\mathscr{L}^{(o)}} \ln c_{n(l)}\right\}.$$

Introduce the lattice $\tilde{Z}^2$ dual to Z^2. Its points are of the form $(x_1+1/2, x_2+1/2)$ with arbitrary integers x_1, x_2.

Lemma. *Let $\tilde{V}$ be the set of those points of the dual lattice $\tilde{Z}^2$ whose distance from V does not exceed* 1. *Then, for every function $n(l)$ defined on $\mathscr{L}^{(o)}$ and satisfying*

1) $n(-l) = -n(l)$;

2) $\Delta n(x) = \sum_{l=(x,y)\in\mathscr{L}^{(o)}} n(l) = 0$,

it is possible to find a function $m(\tilde{x})$ defined on $\tilde{V}$ and satisfying the relation $n(l) = m(\tilde{x}') - m(\tilde{x}'')$, where $\tilde{x}', \tilde{x}''$ are centres of lattice cells of the lattice Z^2 such that l is their common boundary and moreover x' lies to the left and x'' to the right of l.

The function $m(\tilde{x})$ is uniquely defined up to a constant.

Proof. The lemma can be easily proved if cohomology theory is used. However, we are going to give a direct proof. Choose any edge $l\in\mathscr{L}^{(o)}$ and the corresponding left and right centres $\tilde{x}'$ and $\tilde{x}''$. Let $m(\tilde{x}')$ be arbitrary, and put $m(\tilde{x}'') = m(\tilde{x}') - n(l)$. Now we extend the definition of $m(\tilde{x})$ to a further centre whose distance from $\{\tilde{x}', \tilde{x}''\}$ is equal to 1, and so on. By properties 1) and 2) this definition will not depend on the route along which we construct the continuation. Q.e.d.

It follows from the lemma that there exists a one-to-one correspondence between functions $n(l)$ satisfying conditions 1) and 2) of the lemma and functions $m(\tilde{x})$, $\tilde{x}\in\tilde{V}$ for which $\sum_{\tilde{x}\in\tilde{V}} m(\tilde{x}) = 0$. Thus, (1.6) goes over to the final form

$$\Theta = \sum_{\substack{m(\tilde{x}) \\ \Sigma m(\tilde{x})=0 \\ \tilde{x}\in\tilde{V}}} \exp\left\{\frac{1}{2}\sum_{\|\tilde{x}'-\tilde{x}''\|=1} \ln c_{m(\tilde{x}')-m(\tilde{x}'')}\right\}. \tag{1.10}$$

Consider a model where the integer-valued variables $\varphi(\tilde{x})$ are defined on the dual lattice Z^2 and $\tilde{\mathscr{U}}(\varphi(\tilde{x}'), \varphi(\tilde{x}'')) = -\frac{1}{2}\ln c_{\varphi(\tilde{x}')-\varphi(\tilde{x}'')}$. Then (1.10) says that the partition function of our original model is equal to the partition function of this dual model with the Hamiltonian $\mathscr{H}(\varphi) = \sum\tilde{\mathscr{U}}(\varphi(\tilde{x}'), \varphi(\tilde{x}''))$ and with the additional assumption $\sum_{\tilde{x}\in\tilde{V}} \varphi(\tilde{x}) = 0$.

If the potential in the original model is the negative logarithm of the θ-function, then $\mathscr{U}(\varphi(\tilde{x}'), \varphi(\tilde{x}'')) = c(\varphi(\tilde{x}') - \varphi(\tilde{x}''))^2$. The constructed model is sometimes called the integer-valued Ising model. For χ_V it is natural to take the measure whose weight for any function $\varphi(\tilde{x})$ is equal to 1. Our constructed model can be considered as dual to the original X–Y model. It is also possible to get a correspondence between the finite-dimensional distributions of the original and the dual models.

5. Existence of limit Gibbs distributions

The first problem arising in connection with the definitions of Sections 1–3 is the existence problem: for which Hamiltonians does there exist at least one limit Gibbs distribution? We start with the first simple result connected with this problem.

Let Φ be a compact metric space. Then, by Tikhonov's theorem, Ω is also a compact metrix space in the product topology. Suppose that, for any finite subset $V \in Z^d$, the function $\mathscr{H}_V(\varphi) = \mathscr{H}(\varphi(V)) + \mathscr{H}(\varphi(V)|\varphi(Z^d - V))$ is continuous on Ω and the measure χ (see definition 1.2) is finite.

Theorem 1.2. *Under the described conditions for the Hamiltonian $\mathscr{H}$, there exists at least one limit Gibbs distribution.*

Proof. Choose an arbitrary increasing sequence of finite subsets $V_1 \subset V_2 \subset \ldots \ldots \subset V_k \subset V_{k+1} \subset \ldots, \bigcup_k V_k = Z^d$ and an arbitrary sequence of boundary configurations $\varphi(Z^d - V_k)$, $k = 1, 2, \ldots$. The sequences V_k and $\varphi(Z^d - V_k)$ determine a sequence of probability measures P_k on Ω:

$$P_k(\varphi(Z^d - V_k)) = 1,$$

$$P_k(d\varphi(V_k)|\varphi(Z^d - V_k)) = p(\varphi(V_k)|\varphi(Z^d - V_k)) \prod_{x \in V_k} d\chi(\varphi(x)),$$

where p is given by equation (1.2). By the compactness of Ω, the sequence P_k contains a subsequence converging weakly to a distribution P on Ω. For the sake of simplicity, we suppose that P_k itself converges weakly to P. We show that P is a limit Gibbs distribution for the Hamiltonian $\mathscr{H}$.

It is sufficient to check that, for arbitrary disjoint subsets V and W and any pair of continuous bounded functions $f(\varphi(V))$, $g(\varphi(W))$, the following equation holds

$$\int_\Omega f(\varphi(V))\, g(\varphi(W))\, dP(\varphi) =$$

$$= \int g(\varphi(W))\, dP(\varphi) \cdot \left[\int f(\varphi(V))\, p(\varphi(V)|\varphi(Z^d - V)) \prod_{x \in V} d\chi(\varphi(x))\right].$$

The analogous relation holds true for the distribution P_k if k is so large that $V_k \supset V$:

$$\int_\Omega f(\varphi(V))\, g(\varphi(W))\, dP_k(\varphi) =$$

$$= \int g(\varphi(W))\, dP_k(\varphi) \left[\int f(\varphi(V))\, p(\varphi(V)|\varphi(Z^d - V)) \prod_{x \in V} d\chi(\varphi(x))\right].$$

The integral $\int f(\varphi(V))p(\varphi(V)|\varphi(Z^d-V))\prod_{x\in V}d\chi(\varphi(x))$ is a continuous function of $\varphi(Z^d-V)$ by the continuity of $\mathscr{H}_V(\varphi)$. Thus, by taking the limit $k\to\infty$ on both sides of the last relation, we obtain the previous equation. Q.e.d.

Now we turn to an existence theorem for limit Gibbs distributions in the case when the range space Φ of the variables $\varphi(x)$, $x\in Z^d$, is a complete, separable metric space. Then, for any finite $V\subset Z^d$, the space $\Omega(V)$ of configurations $\varphi(V)=\{\varphi(x), x\in V\}$ is obviously a complete, separable metric space, too.

Definition 1.4. *Let $\mathscr{M}$ be a complete, separable metric space. The continuous function h defined on $\mathscr{M}$ is called compact if, for any real t, the set $\{m:h(m)\leqq t\}$ is compact in $\mathscr{M}$.*

Definition 1.5. *The family of probability measures $\{P\}$ on $\mathscr{M}$ is called relatively compact if from any sequence $\{P_n\}\subset\{P\}$ it is possible to select a weakly convergent subsequence.*

The following theorem rewrites a well-known criterion of relative compactness of Prokhorov.

Prokhorov's theorem. *A family $\{P\}$ of probability measures is relatively compact if there exists a non-negative, compact function $h(m)$, $m\in\mathscr{M}$, and a constant C such that $\int_{\mathscr{M}} h(m)\,dP(m)<C$ for every $P\in\{P\}$.*

Return to the space Ω of infinite configurations $\varphi=\{\varphi(x), x\in Z^d\}$ and fix an increasing sequence of finite subsets $V_k\subset Z^d$, $V_k\subset V_{k+1}$, $\bigcup_{k=1}^{\infty}V_k=Z^d$. If Φ is a complete separable metric space, then Ω has a complete, separable metrization, too; namely, for arbitrary $\bar\varphi=\{\bar\varphi(x), x\in Z^d\}$, $\bar{\bar\varphi}=\{\bar{\bar\varphi}(x), x\in Z^d\}\in\Omega$, we can – for example – put

$$d(\bar\varphi,\bar{\bar\varphi}) = \sum_{x\in Z^d}\frac{1}{2^{\|x\|}}\frac{d(\bar\varphi(x),\bar{\bar\varphi}(x))}{1+d(\bar\varphi(x),\bar{\bar\varphi}(x))}.$$

Thus the weak convergence of probability measures defined on Ω makes sense. It is easy to understand that a sequence $\{P_n\}$ of probability measures on Ω converges weakly if and only if, for every V_k, $k\geqq 1$, the restriction $(P_n|\Omega(V_k))$ is a weakly convergent sequence of probability measures on the complete separable metric space $\Omega(V_k)$. Set $P^{(r)}=\lim_{n\to\infty}(P_n|\Omega(V_r))$. Obviously, the measures $P^{(r)}$ are consistent and, by Kolmogorov's theorem, they determine a unique probability measure $\bar P$ given on the σ-algebra γ of the space Ω such that the restriction of $\bar P$ to any $\Omega(V_r)$ is just $P^{(r)}$. It is also clear that the limit P does not depend on the choice of the sequence $\{V_k\}$.

Lemma. *The family of probability measures $\{P\}$ on Ω is relatively compact if, for every $x\in Z^d$, there is a constant $C_x>0$ and a non-negative, compact function $h_x(\varphi)$*

defined on Φ such that, for every $P \in \{P\}$,

$$\int_\Omega h_x(\varphi(x))\,dP < C_x.$$

Proof. In fact, the function $h_{V_k}(\varphi(V_k)) = \max_{x \in V_k} h_x(\varphi(x))$ is non-negative and compact on the space $\Omega(V_k)$ and $\int h_{V_k}(\varphi(V_k))\,dP \leqq \sum_{x \in V_k} C_x$. The lemma follows from Prokhorov's theorem. Q.e.d.

Now we formulate the conditions of the existence theorem for limit Gibbs distributions.

Suppose that, for the Hamiltonian $\mathscr{H}$, it is possible to find:

1) a compact, non-negative function h on the space Φ;

2) a family of numbers $\{c(x, y),\ x \in Z^d, y \in Z^d\}$ with $\sum_{y \in Z^d} |c(x, y)| < c < 1$ for every $x \in Z^d$, and a number $k > 0$ such that, for every finite $V \subset Z^d$ and arbitrary boundary condition $\varphi(Z^d - V)$ satisfying $\max_{x \in V} \sum_{y \in Z^d} |c(x, y)|\, h(\varphi(y)) < \infty$, the conditional Gibbs distribution $p(\cdot\,|\varphi(Z^d - V))$ (see definition 1.2) exists and satisfies

$$\int h(\varphi(x))\, p(\varphi(x)|\varphi(Z^d - \{x\}))\, d\chi(\varphi(x)) \leqq \leqq K + \sum_{y:\, y \neq x} c(x, y)\, h(\varphi(y)). \tag{1.11}$$

Theorem 1.3. (Dobrushin). *Suppose that in addition to conditions 1) and 2) formulated above the following condition is fulfilled:*

3) for every $x \in Z^d$ and arbitrary bounded, continuous function $g(\varphi)$ on Φ, there exists a sequence of finite subsets $W_n \subset Z^d$, $\bigcup_n W_n = Z^d - \{x\}$, a set of non-negative numbers $d^{(n)}(x, y)$, $y \neq x$, with $\sum_{y \in Z^d} d^{(n)}(x, y) \leqq D^{(n)} \to 0$ as $n \to \infty$, and a sequence of bounded, continuous functions $f_n(\varphi(W_n))$ such that

$$\left|\int g(\varphi(x))\, p(\varphi(x)|\varphi(Z^d - \{x\})) \cdot d\chi(\varphi(x)) - f_n(\varphi(W_n))\right| \leqq$$

$$\leqq D^{(n)} + \sum_{y \neq x} d^{(n)}(x, y)\, h(\varphi(y))$$

holds for arbitrary boundary condition $\varphi(Z^d - \{x\})$ satisfying $\sum_{y \neq x} |c(x, y)|\, h(\varphi(y)) < \infty$.

Then, for the Hamiltonian $\mathscr{H}$, the set of limit Gibbs distributions is not empty.

Proof. Fix an increasing sequence of subsets $V_n \subset Z^d$ and of boundary configurations $\bar{\varphi}(Z^d - V_n)$ such that, for some constant A, $h(\varphi(y)) \leqq A$, $y \in Z^d - V_n$. Define a sequence of probability measures P_n on Ω as follows

$$P_n(\bar{\varphi}(Z^d - V_n)) = 1,$$

$$P(d\varphi(V_n)|\bar{\varphi}(Z^d - V_n)) = p(\varphi(V_n)|\bar{\varphi}(Z^d - V_n))\, \Pi_{y \in V_n} d\chi(\varphi(y))$$

(see (1.3)). We show first that P_n is a weakly compact sequence. This follows from our previous theorem and the following lemma.

Lemma.

$$\sup_n \int h(\varphi(x))\, dP_n \leqq \frac{K}{1-c} + \frac{A}{1-c} = K_0$$

Proof of the Lemma. We prove that, for every n,

$$m = \max_{x\in V_n} \int h(\varphi(x))\, p(\varphi(V_n)|\bar{\varphi}(Z^d - V_n)) \prod_{y\in V_n} d\chi(\varphi(y)) \leqq K_0.$$

If we knew that $m<\infty$, then by (1.11) we would have

$$\int h(\varphi(x))\, p(\varphi(x)|\bar{\varphi}(V_n - x),\, \bar{\varphi}(Z^d - V_n))\, d\chi(\varphi(x)) \leqq$$
$$\leqq K + \sum c(x, y) h(y) + cA.$$

Fix x for which the maximum m is attained. Then, by integrating both sides of this inequality with respect to the distribution P_n, we get that the number m satisfies the inequality

$$m \leqq K + \sum c(x, y) m + cA = K + mc + cA$$

or

$$m \leqq (K + cA)(1-c)^{-1} \leqq K_0.$$

Since we do not know yet that $m<\infty$, this argument should be somewhat modified. We show that any non-negative function $h(\varphi(x))$ satisfying $\int h(\varphi(x)) p(\varphi(x)|\varphi(V_n - x),\, \varphi(Z^d - V_n))\, d\chi(\varphi(x)) \leqq \sum_{y\neq x} c(x, y) h(y) + K$ is P_n-integrable and its integral with respect to P_n does not exceed $(K+A)(1-c)^{-1}$.

For any $\varepsilon>0$, we introduce the set $\mathscr{E}_\varepsilon$ of measurable functions $e(\varphi(V_n))$ with $0 \leqq e(\varphi(V_n)) \leqq 1$, $\int e(\varphi(V_n))\, dP_n \geqq 1-\varepsilon$. Set

$$G_\varepsilon = \inf_{e\in\mathscr{E}_\varepsilon} \max_{x\in V_n} \int h(\varphi(x)) e(\varphi(V_n))\, dP_n.$$

Clearly $G_\varepsilon<\infty$. We will show that $G_\varepsilon \leqq (K+A)(1-c)^{-1}$, which will imply (by Fatou's lemma) the required assertion.

Fix $\delta>0$ and choose a function $e_0 \in \mathscr{E}_\varepsilon$ such that

$$\max_{x\in V_n} \int h(\varphi(x)) e_0(\varphi(V_n))\, dP_n \leqq G_\varepsilon(1+\delta)$$

and, for at least one point $x_0 \in V_n$,

$$\int h(\varphi(x_0))\, e_0(\varphi(V_n))\, dP_n \geqq G_\varepsilon(1-\delta).$$

e_0 can be chosen in such a way that the cardinality of points $x\in V_n$ for which the last inequality is fulfilled be minimal. This is possible because V_n is a finite set.

Set $\quad \tilde{e}_0(\varphi(V_n)) = \tilde{e}_0(\varphi(V_n - x_0)) = \int e_0(\varphi(V_n)) \cdot p(\varphi(x_0)|\varphi(V_n - x_0),$

$$\varphi(Z^d - V_n))\, d\chi(\varphi(x_0)).$$

Obviously $\tilde{e}_0 \in \mathscr{E}_\varepsilon$ and, for every $x \in V_n - \{x_0\}$,

$$\int h(\varphi(x))\, \tilde{e}_0\, dP_n = \int h(\varphi(x))\, e_0\, dP_n \leqq G_\varepsilon(1+\delta).$$

Thus, from the minimality of e_0, it follows that

$$\int \tilde{e}_0(\varphi(V_n))\, h(\varphi(x_0))\, dP_n \geqq G_\varepsilon(1-\delta).$$

Now, from condition 2) of the theorem

$$G_\varepsilon(1-\delta) \leqq \int \tilde{e}_0(\varphi(V_n))\, h(\varphi(x_0))\, dP_n =$$

$$= \int \tilde{e}_0(\varphi(V_n - x_0))\, dP_n(\varphi(V_n - x_0)) \cdot \int h(\varphi(x_0))\, p(\varphi(x_0)|\varphi(V_n - x_0), \varphi(Z^d - V_n)) \cdot$$

$$\cdot\, d\chi(\varphi(x_0)) \leqq \int \tilde{e}_0(\varphi(V_n - x_0)) \left(\sum\nolimits_{y\in V_n - x_0} c(x_0, y)\, h(\varphi(y)) + cA\right) dP_n \leqq$$

$$\leqq K + \sum\nolimits_y c(x_0, y)\, G_\varepsilon(1+\delta) + cA \leqq K + cG_\varepsilon(1+\delta) + cA$$

or

$$G_\varepsilon(1 - \delta - c(1+\delta)) \leqq A + K.$$

But δ is arbitrary and consequently

$$G_\varepsilon \leqq (K+A)(1-c)^{-1}.$$

Hence the lemma.

Thus the sequence P is relatively compact. Without any loss of generality, we can suppose that P_n converges weakly to a limit $\bar{P}$, $\bar{P} = \lim_{n\to\infty} P_n$. We show that $\bar{P}$ is a limit Gibbs distribution for the Hamiltonian $\mathscr{H}$.

First, let us observe that the weak convergence of P_n to $\bar{P}$ and the continuity of h imply $\int h(\varphi(x))\, d\bar{P}(\varphi) \leqq K_0$ for every $x \in Z^d$. Thus the expected value of the function $\sum_{y\in Z^d} |c(x, y)| h(\varphi(y))$ is finite and consequently the series $\sum c(x, y)\varphi(y)$ converges with $\bar{P}$-probability 1. This implies that, for every V, the $\bar{P}$-probability of boundary conditions $\varphi(Z^d - V)$ such that $\sum_{y\in Z^d - V} |c(x, y)| h(\varphi(y)) < \infty$ is 1. We show that the conditional distribution on $\Omega(V)$ is a conditional Gibbs distribution. By what has been said we can restrict our attention to boundary conditions where the series $\sum_{y\in Z^d - V} |c(x, y)| h(\varphi(y))$ converges absolutely.

Further, we can prove this statement for subsets V consisting of a single point. This follows from a well-known fact of measure theory: let $(\mathscr{M}, \gamma, \chi)$ be a direct product of measure spaces $(\mathscr{M}_i, \gamma_i, \chi_i)$, i.e. $(\mathscr{M}, \gamma, \chi) = \prod_{i=1}^n (\mathscr{M}_i, \gamma_i, \chi_i)$. Then any probability measure P on $\mathscr{M}$ which is absolutely continuous with respect to the product measure χ is uniquely determined by its conditional probabilities $P(\cdot|m_1, m_2, \dots, m_{i-1}, m_{i+1}, \dots, m_n)$ defined on γ_i for almost every $(m_1, m_2, \dots$ $\dots, m_{i-1}, m_{i+1}, \dots, m_n)$; $i = 1, \dots, n$, $m_i \in \mathscr{M}_i$.

Let us fix $x \in Z^d$, $V \subset Z^d - x$ and the bounded, continuous functions u, v defined on Φ and $\Omega(V)$ respectively. It is sufficient to prove that

$$\int v(\varphi(V))\, u(\varphi(x))\, d\bar{P}(\varphi) = \tag{1.12}$$

$$= \int_\Omega v(\varphi(V))\, d\bar{P}(\varphi) \int u(\varphi(x))\, p(\varphi(x)|\varphi(Z^d - x))\, d\chi(\varphi(x)).$$

Obviously, the following equation holds true if $V \subset V_n$

$$\int v(\varphi(V))u(\varphi(x))\,dP_n(\varphi) = \tag{1.13}$$

$$= \int_\Omega v(\varphi(V))\,dP_n(\varphi) \int u(\varphi(x))\,p(\varphi(x)|\varphi(Z^d - x))\,d\chi(\varphi(x)).$$

The left-hand side of (1.13) converges to the left-hand side of (1.12) as $n \to \infty$ by the definition of weak convergence.

To handle the right-hand side of (1.13), we shall make use of condition 3) of the theorem. By applying it to the function u, we get

$$\left|\int u(\varphi(x))\,p(\varphi(x)|\varphi(Z^d - x))\,d\chi(\varphi(x)) - f_k(\varphi(W_k))\right| \leqq \tag{1.14}$$

$$\leqq D^{(k)} + \sum_{y \neq x} d^{(k)}(x, y)\,h(\varphi(y)).$$

Then

$$\int u(\varphi(x))\,p(\varphi(x)|\varphi(Z^d - x))\,d\chi(\varphi(x)) = f_k(\varphi(W_k)) + \varepsilon_k(\varphi)$$

where $|\varepsilon_k| \leqq D^{(k)} + \sum_{y \neq x} d^{(k)}(x, y)\,h(\varphi(y))$. Using this in (1.13), we obtain

$$\int v(\varphi(V))u(\varphi(x))\,dP_n(\varphi) = \int v(\varphi(V))(f_k(\varphi(W_k)) + \varepsilon_k(\varphi))\,dP_n(\varphi).$$

Taking the limit $n \to \infty$, we find

$$\int v(\varphi(V)) f_k(\varphi(W_k))\,d\bar{P} - D^{(k)}(1 + K_0) \max_{\varphi(V)} |v(\varphi(V))| \leqq$$

$$\leqq \int v(\varphi(V))u(\varphi(x))\,d\bar{P} \leqq$$

$$\leqq \int v(\varphi(V)) f_k(\varphi(W_k))\,d\bar{P} + D^{(k)}(1 + K_0) \max_{\varphi(V)} |v(\varphi(V))|.$$

Inequality (1.14) and the lemma imply that, if $k \to \infty$, then f_k converges to $\int u(\varphi(x))p(\varphi(x)|\varphi(Z^d - x))\,d\chi(\varphi(x))$ in the space $\mathscr{L}^1(\Omega, \gamma, \bar{P})$. Thus we can take the limit $k \to \infty$ in the last inequality, that results in (1.12). Q.e.d.

Remark 1. If Φ is compact, we can choose $h \equiv 0$.

Remark 2. If $p(\varphi(x)|\varphi(Z^d - x))$ is defined everywhere and is continuous in the Tikhonov topology, then condition 3) of the theorem is fulfilled with $d^{(n)}(x, y) \equiv 0$.

Remark 3. If the Hamiltonian $\mathscr{H}$ is translation invariant, then, for every limit Gibbs distribution $\bar{P}$, its translate $(T^x)^*\bar{P}$ is also a limit Gibbs distribution. By modifying our construction, it is also possible to get a translationally invariant $\bar{P}$.

Remark 4. The methods of the proof of existence of limit Gibbs distributions can be developed further to obtain a general description of the set $G(\mathscr{H})$ of limit Gibbs distributions for a given Hamiltonian $\mathscr{H}$. Let $G_V^0(\mathscr{H})$ denote the set of finite convex combinations of conditional Gibbs distributions in a finite volume V with different boundary conditions, and let $G_V(\mathscr{H})$ be the closure of $G_V^0(\mathscr{H})$ in the weak topology of probability distributions. Under some natural regularity conditions on the interaction, $G(\mathscr{H})$ has the following structure.

(i) $P \in G(\mathscr{H})$ if and only if there exist sequences $V_n \subset Z^d$ and $P_n \in G^0_{V_n}(\mathscr{H})$ such that $V_n \to \infty$ and $P = \lim P_n$. This assertion motivates our terminology that elements of $G(\mathscr{H})$ are called limit Gibbs distributions.

(ii) $G(\mathscr{H})$ is a nonempty, convex, compact set of probability distributions, and $G(\mathscr{H}) = \bigcap G_V(\mathscr{H})$, where the intersection is over all finite $V \subset Z^d$. The extremal points of $G(\mathscr{H})$ are called indecomposable limit Gibbs distributions.

(iii) $P \in G(\mathscr{H})$ is indecomposable if and only if P is regular in the sense that for every event A we have

$$\lim_n |P(AB_n) - P(A)P(B_n)| = 0$$

whenever B_n belongs to the σ-algebra generated by the variables $\{\varphi(x) | x \in Z^d - V_n\}$ and $V_n \to \infty$.

(iv) Different extremal points of $G(\mathscr{H})$ are mutually singular, so that $G(\mathscr{H})$ is a Choquet simplex.

(v) If $\mathscr{H}$ is Z_0-periodic with respect to a subgroup $Z_0 \subset Z^d$ of finite index, then the set $G_0(\mathscr{H})$ of Z_0-periodic limit Gibbs distributions is again a (nonempty) Choquet simplex. Extremal points of $G_0(\mathscr{H})$ are exactly the Z_0-ergodic limit Gibbs distributions.

The above statements can be easily verified for example if Φ is a compact metric space and each conditional Gibbs distribution is a continuous function of the boundary condition.

Example. Existence theorem for limit Gibbs distributions for lattice models of quantum field theory.

In these models $\Phi = R^1$ and the Hamiltonian is (see Section 4, Example 4):

$$\mathscr{H} = \mathscr{H}_0 + \mathscr{H}^{(\mathrm{int})} = \sum_{\|x-y\|=1} (\varphi(x) - \varphi(y))^2 + + m_0^2 \sum_{x \in Z^2} \varphi^2(x) + \sum_{x \in Z^2} Q(\varphi(x)), \tag{1.15}$$

where $Q(t) = \sum_{p=1}^{2r} a_p t^p$, $a_{2r} > 0$, $r \geqq 1$ and

$$\mathscr{H}_0 = \sum_{\|x-y\|=1} (\varphi(x) - \varphi(y))^2 + m_0^2 \sum_{x \in Z^2} \varphi^2(x).$$

The Hamiltonian (1.15) is binary, translation invariant with interaction radius $R=1$. For measure χ we choose the Lebesgue measure. The conditional distributions in any finite domain V are defined for all possible boundary conditions. We show using Dobrushin's theorem that there exists at least one limit Gibbs distribution for the Hamiltonian (1.15).

For $r=1$, we have a quadratic Hamiltonian, which determines a stationary Gaussian distribution (see Section 4, Example 3). Therefore we suppose further on that $r>1$. For auxiliary function we choose $h(\varphi) = \mathrm{ch}\, \varphi$. Then the problem leads to finding constants K, $c(x, y)$ with $\sum_y |c(x, y)| < 1$ such that

$$(1.16)\qquad \frac{\int \operatorname{ch} \varphi(x) \exp\{2\varphi(x) \sum_{y:\, \|x-y\|=1} \varphi(y) - m_0^2 \varphi^2(x) - Q(\varphi(x))\}\, d\varphi(x)}{\int \exp\{2\varphi(x) \sum_{y:\, \|x-y\|=1} \varphi(y) - m_0^2 \varphi^2(x) - Q(\varphi(x))\, d\varphi\}(x)} \leqq$$

$$\leqq K + \sum_{y \neq x} c(x, y) \operatorname{ch} \varphi(y).$$

Put $c(x, y)$ equal to $(4d)^{-1}$ if $\|x-y\|=1$, and zero otherwise. Clearly, $\sum_y |c(x, y)| = 1/2$. It remains to find K. We show that the construction of a required K follows by some formulae that can be obtained by applying the well-known asymptotic Laplace method.

Consider the integral

$$F(a) = \int_{-\infty}^{\infty} \exp\{a\varphi - P(\varphi)\}\, d\varphi,$$

where $P(\varphi) = \sum_{p=1}^{2r} c_p \varphi^p$, $c_{2r} > 0$.

Lemma. *For $|a| \to \infty$, we have the asymptotic relation*

$$F(a) \sim \exp\{af(a) - P(f(a))\} \left(\frac{2\pi}{P''(f(a))}\right)^{1/2},$$

where $f(a)$ is the minimum of the function $[at - P(t)]$, which is unique for large values of $|a|$.

We shall prove the lemma a bit later. Plug into (1.16)

$$b = 2 \sum_{y:\, \|x-y\|=1} \varphi(y), \quad P(\varphi) = Q(\varphi) + m_0^2 \varphi^2$$

and

$$D(b) = \frac{\int_{-\infty}^{\infty} \operatorname{ch} \varphi \exp(b\varphi - P(\varphi))\, d\varphi}{\int_{-\infty}^{\infty} \exp(b\varphi - P(\varphi))\, d\varphi} =$$

$$= \frac{1}{2} \frac{\int_{-\infty}^{\infty} \exp((b+1)\varphi - P(\varphi))\, d\varphi}{\int_{-\infty}^{\infty} \exp(b\varphi - P(\varphi))\, d\varphi} +$$

$$+ \frac{1}{2} \frac{\int_{-\infty}^{\infty} \exp((b-1)\varphi - P(\varphi))\, d\varphi}{\int_{-\infty}^{\infty} \exp(b\varphi - P(\varphi))\, d\varphi} = D_1(b) + D_2(b).$$

Then, by the lemma,

$$D_1(b) \sim \frac{1}{2} \exp\{(b+1) f(b+1) - P'(f(b+1)) - bf(b) + P'(f(b))\},$$

$$D_2(b) \sim \frac{1}{2} \exp\{(b-1) f(b-1) - P'(f(b-1)) - bf(b) + P'(f(b))\}.$$

Set $L(a) = af(a) - Q(f(a))$ for $|a|$ large. Then $L'(a) = f(a)$. Suppose $b > 0$. (The case

$b<0$ is analogous.) Since $f(a)\to\infty$ as $a\to\infty$, we have

$$D_1(b)\sim\frac{1}{2}\exp\big(L(b+1)-L(b)\big)\to\infty \quad \text{if} \quad b\to\infty,$$

$$D_2(b)\sim\frac{1}{2}\exp\big(L(b-1)-L(b)\big)\to 0 \quad \text{if} \quad b\to\infty,$$

and by Lagrange's theorem, for some $\theta(b)$, $0<\theta(b)<1$,

$$D(b)\sim D_1(b)\sim\frac{1}{2}\exp\big(L(b+1)-L(b)\big)=$$

$$=\frac{1}{2}\exp\big(f(b+\theta(b))\big).$$

It is easily seen that $f(b)\sim\text{const}\cdot b^{(2r-1)^{-1}}$ for $b\to\infty$ and, consequently, $f\big(b+\theta(b)\big)-f(b)\to 0$ as $b\to\infty$. Thus, we obtain finally

$$D(b)\sim\frac{1}{2}\exp|f(b)| \quad \text{if} \quad b\to\infty.$$

Choose A in such a way that for $|b|>A$, $D(b)<\exp|f(b)|$ and $\exp|f(b)|<\frac{1}{2}\operatorname{ch}(b/4d)$. By putting $K=\max_{|b|\leqq a}D(b)$ we have for $|b|\leqq A$,

$$D(b)\leqq K\leqq K+\sum\nolimits_{\|x-y\|=1}c(x,y)\operatorname{ch}\varphi(y),$$

and for $|b|>A$,

$$D(b)<\exp|f(b)|<\frac{1}{2}\operatorname{ch}\left(\frac{b}{4d}\right)=$$

$$=\frac{1}{2}\operatorname{ch}\left(\frac{1}{2d}\sum\nolimits_{\|x-y\|=1}\varphi(y)\right)\leqq\frac{1}{4d}\sum\nolimits_{\|x-y\|=1}\operatorname{ch}\varphi(y)\leqq$$

$$\leqq K+\sum c(x,y)h\big(\varphi(y)\big),$$

using Jensen's inequality for convex functions.

Proof of the Lemma. Setting $\varphi=f(a)+t$, we get

$$F(a)=\exp\{af(a)-P\big(f(a)\big)\}\int_{-\infty}^{\infty}\exp\left[-\frac{P''(f(a))}{2!}t^2-\dots-\frac{P^{(2r)}(f(a))}{(2r!)}t^{2r}\right]dt.$$

Divide the integration into two parts: for $|t|<a^{(3/2-2r)(2r-1)^{-1}}$ and for $|t|>a^{(3/2-2r)(2r-1)^{-1}}$. Since $f(a)=O(a^{(2r-1)^{-1}})$ we have $P^{(k)}\big(f(a)\big)=O(a^{(2r-k)(2r-1)^{-1}})$ and, in the first domain of integration, $|P^{(k)}(f(a))t^{2k}|\leqq\text{const}\cdot a^{(3/2-k)(2r-1)^{-1}}$. If $k>1$, then the last expression tends to 0. Now it follows easily that, in the first domain of integration the integral is equivalent to

$$\int_{-\infty}^{\infty} \exp\left\{-\frac{P''(f(a))}{2}t^2\right\}dt = \left(\frac{2\pi}{P''(f(a))}\right)^{1/2}.$$

Further, in the second domain of integration, the function

$$\frac{P''(f(a))}{2}(1-\alpha)t^2+\ldots+\frac{P^{(2r)}(f(a))}{(2r)!}t^{2r}$$

is non-negative for some α, $0<\alpha<1$. This implies immediately that the integral over this domain is of smaller order than the one in the first domain. Q.e.d.

6. Limit Gibbs distributions for continuous fields and for point fields

There are two more classes of random fields where the construction of limit Gibbs distributions plays an important role. We shall only touch on these classes and in the forthcoming chapters they will scarcely be mentioned. We emphasize that, in spite of their extraordinary importance, the study of these classes of fields is not a completely explored domain of the theory of random fields.

Start with d-dimensional random fields with continuous time. Let Ω be a topological vector space of real-valued functions $\varphi(x)=\varphi(x_1, \ldots, x_d)$ defined on R^d. Let μ_0 be a probability distribution on the Borel σ-algebra of subsets of Ω. μ_0 is the free field. It can be, for example, a Gaussian distribution. Choose a function $\mathcal{U}(\varphi)$ such that, for arbitrary open domain $V\subset R^d$, $\int \exp[-\int_V \mathcal{U}(\varphi(x))\,dx]\,d\mu_0(\varphi)<\infty$. Then, by fixing the values of the function φ outside V, i.e. the configuration $\varphi(R^d-V)$, we can construct conditional distributions $\mu_0(\cdot\,|\varphi(R^d-V))$ on configurations $\varphi(V)$ inside the domain V, and we can introduce conditional distributions $P(\cdot\,|\varphi(R^d-V))$, absolutely continuous with respect to $\mu_0(\cdot\,|\varphi(R^d-V))$, by prescribing their densities as follows

$$\frac{dP(\varphi(V)|\varphi(R^d-V))}{d\mu_0(\varphi(V)|\varphi(R^d-V))} = \Theta^{-1}(\varphi(R^d-V))e^{-\int_V \mathcal{U}(\varphi(x))\,dx}, \tag{1.17}$$

where

$$\Theta(\varphi(R^d-V)) = \int \exp\left\{-\int_V \mathcal{U}(\varphi(x))\,dx\right\} d\mu_0(\varphi(V)|\varphi(R^d-V)).$$

Definition 1.6. *The measure P is called a limit Gibbs distribution if, for almost every (with respect to the distribution P) condition $\varphi(R^d-V)$, the conditional distribution induced by P on the realizations $\varphi(V)$ has the form* (1.17).

The given definition permits a natural generalization to the case when the probability distribution μ_0 corresponds to a generalized random field. Then there appears one of the most important examples of limit Gibbs distributions arising in two-dimensional quantum field theory. Namely, let μ_0 be the two-dimensional stationary, symmetric Gaussian field with Hamiltonian

$$\mathscr{H} = \int [(\nabla\varphi, \nabla\varphi)+m_0^2\varphi^2]\, dx_1\, dx_2 =$$

$$= \int \left[\left(\frac{\partial\varphi}{\partial x_1}\right)^2 + \left(\frac{\partial\varphi}{\partial x_2}\right)^2 + m_0^2\varphi^2\right] dx_1\, dx_2 .$$

This random field can only be determined as a generalized stationary Gaussian field whose correlation operator is given by the Fourier transform of the function $(\lambda_1^2+\lambda_2^2+m_0^2)^{-1}$.

Consider, for an arbitrary domain V, the random variable

$$\int_V \sum_{k=1}^{2r} a_k :\varphi^k(x_1, x_2): dx_1 dx_2 = \mathscr{H}_V^{(\text{int})}(\varphi).$$

This random variable can be determined either by taking the limit of lattice models of quantum field theory, described above, or by the aid of multiple stochastic integrals of the Hermite–Ito type (cf. [19], [90]). In fact, if

$$\varphi(x_1, x_2) = \int e^{i(\lambda_1 x_1+\lambda_2 x_2)}\, d\Phi(\lambda_1, \lambda_2)$$

is the Fourier expansion of a generalized random field, then

$$\mathscr{H}_V^{(\text{int})}(\varphi) = \int_V :\varphi^k(x_1, x_2): dx_1\, dx_2 =$$

$$= \int d\Phi(\lambda^{(1)}) \dots d\Phi(\lambda^{(k)}) \int_V S\left(e^{2\pi i (x, \Sigma_{p=1}^k \lambda^{(p)})}\right) dx,$$

where $S(\cdot)$ is the symmetrization of the function in brackets and the integral on the right-hand side is a multiple stochastic Ito integral.

Nelson [103] has shown that, if $a_{2r}>0$, then $E \exp\{-\mathscr{H}_V^{(\text{int})}(\varphi)\}<\infty$ for domains V with sufficiently nice boundary. Consequently, the construction of conditional Gibbs distributions is possible for such domains. It is not hard to see that the free field μ_0 is Markovian in the sense that the conditional distribution in the domain V depends only on the realization of the field in an arbitrary small neighbourhood of the boundary. Then the conditional Gibbs distribution is Markovian in the same sense, and the same is also true for the limit Gibbs distribution. The exact formulation of these statements is far from being simple.

Another class of limit Gibbs distributions arises in the case of random point fields. The most widespread situation is the one when we consider the space Ω consisting of countable subsets φ of R^d, $d \geqq 1$, such that the intersection of φ with any finite subset $V \subset R^d$ is finite. The natural choice for the free-field μ_0 is a Poisson field with parameter λ. In problems of statistical mechanics it is often the case that the conditional Gibbs distribution in the domain V is absolutely continuous with respect to μ_0 with the density

$$\frac{dP}{d\mu_0} = (\Theta)^{-1} \exp\{-\sum_{x',x''\in V} \mathscr{U}(|x'-x''|) - \sum_{\substack{x'\in V \\ x''\in R^d - V}} \mathscr{U}(|x'-x''|)\}.$$

The function $\mathscr{U}(r)$ is called potential of a pair interaction or briefly pair potential. We suppose it is isotropic, i.e. independent of the direction, and this is why we write it as a function of the distance. It is often assumed that $\mathscr{U}(r)=\infty$ if $r \leqq a$ with $a>0$ and decreases monotonically at infinity so rapidly that $\int^{\infty} r^{d-1} \mathscr{U}(r)\, dr<\infty$. Then limit Gibbs distributions are defined in the same way as above. The existence of a limit Gibbs distribution can be proved in the same way as in the compact case. The existence of limit Gibbs distributions can also be shown under weaker assumptions for the convergence of $\mathscr{U}(r)$ to infinity in the neighbourhood of the origin. The examples described here are the basic ones of equilibrium statistical mechanics of continuous systems.

Historical notes and references to Chapter I

1. The general definition of a limit Gibbs distribution appeared in papers of Dobrushin [13–16] and of Lanford and Ruelle [95]. A particular case of this definition was given much earlier in a paper of Bogolubov and Hacet. As to an extended and modernized variant of this paper see the paper of Bogolubov, Petrina and Hacet [8]. We should also mention the papers of Minlos [25, 26] and of Ruelle [109].

For probability-theoretical aspects of the theory of limit Gibbs distributions see the monograph by Preston [107].

2. Gaussian distributions were investigated from the point of view of limit Gibbs distributions in the papers of Rozanov [38], Spitzer [115] and Dobrushin [57]. For an interesting example of a Gaussian field arising in quantum electrodynamics, see Guerra's paper [82].

3. Yang–Mills fields are intensively studied in the literature of quantum field theory; quite a number of papers are devoted to them. Lattice variants of Yang–Mills fields appeared in the papers of Wilson [116] and Polyakov (unpublished). Some mathematical problems of the theory of these fields are treated in the recent survey by Osterwalder and Seiler [104].

4. From the point of view of limit Gibbs distributions, two-dimensional models of quantum field theory are one of the most interesting topics of research, and the literature about them is quite rich. The idea of constructing limit Gibbs distributions which correspond to two-dimensional models of quantum field theory and also the first theorems about their existence belong to Nelson [103]. Deep methods were worked out by Glimm and Jaffe. We only mention their survey papers [73–75], where other references can also be found. In the book by Simon [41] and the paper of Guerra, Rosen and Simon [81], continuous and lattice models of quantum field theory have been studied from the point of view of statistical mechanics and the theory of limit Gibbs fields.

Interesting results in this topic were obtained by Fröhlich, who, in particular, proved the existence of a limit Gibbs distribution for continuous two-dimensional models of quantum field theory under the widest assumptions [63]. The: φ^4:$_3$ model was studied in the deep paper of Glimm and Jaffe [76] and in the paper of Feldmann and Osterwalder [62], based on it.

5. Berezinsky introduced and discussed his duality in his thesis [3].

6. The existence theorems, with the proofs given in section 5, belong to Dobrushin [13–16]. The example of applying this theorem to lattice models of quantum field theory was considered by K. Khanin, a student at Moscow University. He studied a slightly more general case.

Existence theorems of limit Gibbs distributions for models where $\varphi(x)$ takes on unbounded values were considered in the paper of Lebowitz and Presutti [97].

7. For the criterion of Prokhorov, see P. Billingsley: *Convergence of Probability Measures* (Wiley, 1968; Russian translation: Mir, 1977).

8. For a proof of the statements of Remark 4 to Theorem 1.3, see Minlos [25, 26], Dobrushin [13–15], Lanford and Ruelle [95], Preston [107].

CHAPTER II

PHASE DIAGRAMS FOR CLASSICAL LATTICE SYSTEMS PEIERLS'S METHOD OF CONTOURS

1. Introduction

The most fundamental problem in the theory of phase transitions (equilibrium statistical mechanics) is certainly that of giving a description of limit Gibbs distributions. As discussed in Chapter I, limit Gibbs distributions for a Hamiltonian $\mathscr{H}$ are constructed as infinite volume limits of conditional Gibbs distributions in finite volumes with different boundary conditions. Under fairly general assumptions the set $\mathscr{G}(\mathscr{H})$ of limit Gibbs distributions for $\mathscr{H}$ forms a nonempty Choquet simplex of probability measures (see Remark 4 to Theorem 1.3). Therefore the characterization problem of $\mathscr{G}(\mathscr{H})$ reduces to that of its extremal points. Extremal points of the convex set $\mathscr{G}(\mathscr{H})$ are called indecomposable limit Gibbs distributions; statistical properties of pure thermodynamic phases are described just by indecomposable limit Gibbs distributions. Since only translation-invariant or periodic Hamiltonians are considered here, it is quite natural to restrict our attention to the study of indecomposable and translation-invariant or periodic limit Gibbs distributions; these distributions are frequently referred to as pure phases. In the usual formulation of the above problem a one-parameter family $\beta\mathscr{H}$ of Hamiltonians is considered where the parameter $\beta \geqq 0$ is proportional to the inverse temperature, and $\mathscr{H}$ is a given Hamiltonian. It is extremely hard to give a complete description of $\mathscr{G}(\beta\mathscr{H})$. More or less satisfactory results are obtainable only for small values of β, i.e. at high temperatures. Indeed, suppose that the reference measure χ occurring in the definition (1.2) of conditional Gibbs distributions is a probability measure, and $\beta=0$. Then the conditional Gibbs distribution (1.2) does not depend on the boundary condition and the individual spin variables $\varphi(x)$, $x \in V$, are statistically independent with common distribution χ. Therefore the limiting procedure $V \to \infty$ is trivial; the unique limit Gibbs distribution represents a family $\varphi(x)$, $x \in Z^d$, of independent and χ-distributed random variables. Under very general conditions, the point $\beta=0$ is stable in the sense that for small values of β the limit Gibbs distribution is still unique and it corresponds to a family $\varphi(x)$, $x \in Z^d$, of weakly dependent random variables. The exact formulation of these results can be found e.g. in the monograph by Ruelle [39] and in the papers by Dobrushin [13, 14, 15]. Such proofs are usually based on the contractivity of the so-called correlation equations.

The situation is much more interesting in the neighbourhood of $\beta=\infty$. Consider the formal expression const $\cdot e^{-\beta\mathscr{H}}$ of the Gibbs distribution; such a distribution degenerates as $\beta\to\infty$, i.e. it converges to a probability measure concentrated on the set of such configurations where $\mathscr{H}$ attains its minimum. Of course, this assertion has not any definite meaning because in case of infinite configurations, $\mathscr{H}$ is merely a formal infinite sum. Nevertheless, by introducing the concept of ground states, we shall reformulate the heuristic ideas above in a rigorous way. It will turn out that the point $\beta=\infty$ is stable, too. Namely, for Hamiltonians with a finite set of periodic ground states satisfying a Peierls's stability condition (see later) we shall show that, at large values of β, there appear some indecomposable and periodic limit Gibbs distributions called pure phases, and typical configurations of any pure phase are obtainable by small local distortions of one of the ground states. The whole of Chapter II is devoted to a systematic exploration of the deep content behind this intuitive picture.

In order to obtain a complete picture on the coexistence of pure phases, we let $\mathscr{H}$ depend on some parameters $(\mu_1, \mu_2, \ldots, \mu_k)=\mu$, interpreted as the intensity of certain external fields; i.e. we are given a k-parameter family

$$\mathscr{H}_\mu = \mathscr{H}_0+\sum_{i=1}^{k} \mu_i \mathscr{H}_i$$

of Hamiltonians. For small values of μ we shall construct a set $\hat{G}(\beta\mathscr{H}_\mu)$ of indecomposable and periodic limit Gibbs distributions. If β is fixed, then the number $|\hat{G}(\beta\mathscr{H}_\mu)|$ of these pure phases turns out to be a function of μ and the parameter space splits into some maximal connected sets characterized by the property that $|\hat{G}(\beta\mathscr{H}_\mu)|$ is constant on each of them. This decomposition of the parameter space is called the phase diagram at β for the family $\beta\mathscr{H}_\mu$ of Hamiltonians. Under some natural assumptions we shall prove that, at large values of β, the phase diagram depends weakly on β; its topological structure is the same as that of the diagram corresponding to the number of ground states of $\mathscr{H}_\mu$ (phase diagram at $\beta=\infty$). Thus coexistence of two or more pure phases appears together with the degeneration of the ground state of the basic Hamiltonian $\mathscr{H}_0$; i.e. $\mathscr{H}_0$ has at least two ground states in the domain of phase transitions. Let us remark that in the case of Hamiltonians with certain symmetries the degeneration of the ground state is frequently of such a type that the ground states themselves are not invariant under the symmetry group of the Hamiltonian, i.e. the symmetry group acts in a proper way in the space of ground states. In such situations we say that coexistence of phases is associated with a spontaneous breakdown of symmetry. In general, however, two or more pure phases can coexist without any breakdown of symmetry; the degeneration of the ground state is the principal background phenomenon of phase transitions.

The usual physical interpretation of phase transitions in the low-temperature domain is based on the concept of surface tension. To demonstrate these ideas we present now a heuristic treatment of the two-dimensional Ising ferromagnet. The following reasoning has been inspired by several conversations of the author with I. M. Lifshitz. It will be convenient here to formulate the problem in terms of the so-called microcanonical ensemble. Fix the value c of magnetization, $-1 \leqq c \leqq 1$. Then the corresponding microcanonical statistical sum in a finite volume V is given by

$$\Theta(V;\, c, \beta) = \sum_{\varphi(V):\ \sum_{x \in V} \varphi(x) = [c|V|]} \exp\left(-\beta \mathscr{H}^I(\varphi(V))\right)$$

where $\mathscr{H}^I$ is the Hamiltonian of the Ising ferromagnet, i.e. $\Phi = \{-1, +1\}$, $\mathscr{H}^I = \sum_{\|x-y\|=1} (\varphi(x) - \varphi(y))^2$, $\varphi(V)$ denotes the restriction of a configuration φ to V, $|V|$ denotes the number of points in V and $[c|V|]$ is the integer part of $c|V|$. In view of the Van Hove theorem (see [39]), there exists the limit

$$\lim_{V \to \infty} \frac{1}{|V|} \ln \Theta(V;\, c, \beta) = -\beta f(c, \beta);$$

$f(c, \beta)$ is just the Gibbs free energy, it is a convex function of c. We show that at large values of β a horizontal linear segment happens to appear on the graph of f, indicating coexistence of phases. More exactly, $f(c, \infty)$ does not depend on c, it is equal to the common value of the specific energy for any of the ground state configurations $\varphi \equiv +1$ and $\varphi \equiv -1$. Further $f(c, \beta)$ is constant in an interval $[-c^*(\beta), c^*(\beta)]$, $c^*(\beta) \geqq 0$, $\lim_{\beta \to \infty} c^*(\beta) = 1$ and $\lim_{\beta \to \infty} f(c, \beta) = f(c, \infty)$. (See Fig. 2.1.)

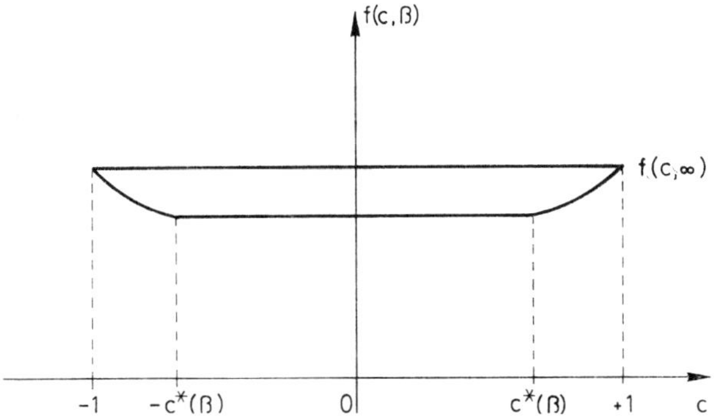

FIGURE 2.1

Suppose now that β is large and c is close to one, then a typical configuration $\varphi(V)$ in a large volume V consists of a "sea" of $+1$ with a few "islands" of -1. More exactly, if Γ denotes the boundary of an island, then the internal side of Γ is occupied by -1, in the remaining part of the interior Int Γ of Γ the configuration is arbitrary, while there are $+1$ everywhere outside the islands, see Fig. 2.2.

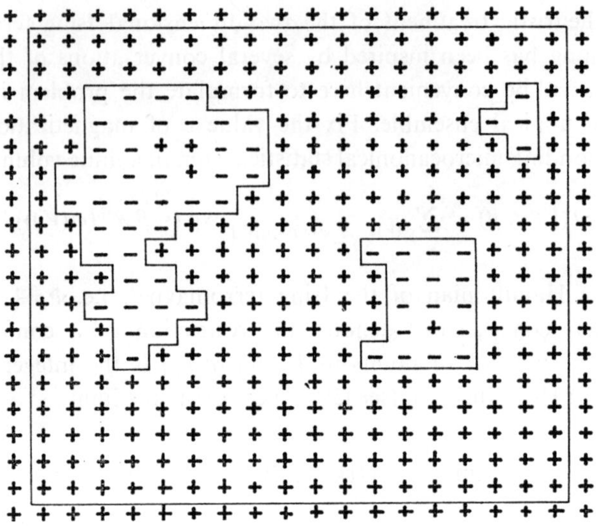

FIGURE 2.2

The boundary Γ of an island of -1 will be called an external contour. Let $\nu_\beta(\Gamma)$ denote the average number of -1 in $\operatorname{Int}\Gamma$ with respect to the conditional Gibbs distribution in $\operatorname{Int}\Gamma$ given that Γ is an external contour, and let $F(\Gamma, \beta)$ denote the logarithm of the corresponding statistical sum. Further, let $\varkappa(\Gamma)$ denote the concentration of external contours of type Γ in $\varphi(V)$; i.e. $\varkappa(\Gamma)|V|$ is the number of external contours Γ' of $\varphi(V)$ such that Γ' is a translate of Γ. Then $2\sum\varkappa(\Gamma)\nu_\beta(\Gamma)=1-c$ and the statistical sum over configurations $\varphi(V)$ with a given set $\{\varkappa(\Gamma)\}$ of concentrations is just

$$K(\{\varkappa(\Gamma)\})\exp\{|V|\sum\varkappa(\Gamma)F(\Gamma,\beta)\},$$

where $K(\{\varkappa(\Gamma)\})$ is the number of such configurations.

To determine the value of $f(c, \beta)$ we have to find the most likely distribution $\{\varkappa(\Gamma)\}$ of external contours. If β is large and c is close to one, then the joint volume of the internal parts of external contours occupies only a small proportion of V, thus we may assume that external contours are selected independently of each other. Indeed, as the density of external contours is low, the possibility of intersections is negligible. In this approximation we have

$$K(\{\varkappa(\Gamma)\})\approx\exp[-|V|\sum\varkappa(\Gamma)\ln\varkappa(\Gamma)],$$

so that

$$f(c,\beta)=-\frac{1}{\beta}\max_{\{\varkappa(\Gamma)\}}\sum\varkappa(\Gamma)[F(\Gamma,\beta)-\ln\varkappa(\Gamma)],$$

where the maximum is taken over such sets $\{\varkappa(\Gamma)\}$ of concentrations that $2\sum\varkappa(\Gamma)v_\beta(\Gamma)=1-c$. Let us remark that $v_\beta(\Gamma)$ is close to $|\text{Int}\,\Gamma|$ if β is large.

Consider now the decomposition $F(\Gamma,\beta)=\alpha_\beta\,|\text{Int}\,\Gamma|-\chi_\beta(\Gamma)|\Gamma|$ of F; here α_β is independent of Γ, it is a function of β only, $\chi_\beta(\Gamma)$ is called the surface tension on Γ. It is possible to prove that $\alpha_\beta\sim e^{-\text{const}\cdot\beta}$ and $\chi_\beta(\Gamma)\sim\text{const}\cdot\beta$ if $\beta\to\infty$. Substituting this expression of F into that of f, the method of Lagrange multipliers results in

$$\varkappa(\Gamma)=\exp\left[-1+\alpha_\beta\,|\text{Int}\,\Gamma|-\mu v_\beta(\Gamma)-\chi_\beta(\Gamma)\,|\Gamma|\right],$$

where the value of μ is determined by the condition $2\sum\varkappa(\Gamma)v_\beta(\Gamma)=1-c$. Hence it follows that $\mu\geqq\alpha_\beta$ and $c=c^*$ for $\mu=\alpha_\beta$. The assumption that β is large is needed for $\sum\varkappa(\Gamma)v_\beta(\Gamma)\ll 1$. Further, $f(c,\beta)$ is a smooth function of c if $-1<c<-c^*$ or $c^*<c<1$, and it is possible to prove that the most probable configurations, i.e. those yielding the main contribution to the microcanonical statistical sum, are such that there exists one large contour Γ^- with

$$\frac{|\text{Int}\,\Gamma^-|}{|V|}(-c^*)+\left(1-\frac{|\text{Int}\,\Gamma^-|}{|V|}\right)c^*=c,$$

and inside Γ^- the concentration of -1 is large while it is small outside Γ^-. In other words, if $-c^*<c<c^*$, then typical configurations are characterized by a concentration and spatial separation of phases.

In a comprehensive and rigorous treatment of these questions the crucial step is to define the concept of surface tension. This means that we have to decompose the logarithm of the statistical sum as $\ln\Theta(\Gamma|\beta)=\alpha_\beta|\text{Int}\,\Gamma|-\chi(\Gamma)|\Gamma|$. As is well known, the surface tension $\chi(\Gamma)$ is not a thermodynamical potential, therefore it is very hard to define $\chi(\Gamma)$ in terms of the Gibbs distribution. Moreover, in case of a second-order phase transition the surface tension is usually infinitely large. Thus it can be defined only for first-order phase transitions. Nevertheless, at least at large values of β some combinatorial methods developed later enable us to define the surface tension for a wide class of lattice systems. An other, less important difficulty is to show that configurations with a spatial separation of phases yield the main contribution to the statistical sum. In the case of the Ising model this assertion has been proved in [28, 29].

In this chapter we shall follow a somewhat different approach towards describing phase diagrams of lattice systems at low temperatures: the framework of the grand canonical ensemble will be used. We shall see that the statistical sum as a function of volume satisfies a certain chain of recurrence equations revealing its fairly peculiar combinatorial structure. In particular, these equations establish an intrinsic relationship of lattice systems to special contour models, introduced in [28, 29] and to contour models with a parameter [34]. Having these tools, it will not be too hard to obtain the information we need on phase diagrams. Also some further details of the above considerations could have been developed on a rigorous level, but we do not proceed to such questions here.

Now we turn to a systematic development of the ideas outlined above. At the end of this section some basic definitions and notations are summarized, as far as possible we follow those of Chapter I.

Throughout this chapter we assume that the individual spin variables $\varphi(x)$, $x\in Z^d$, take values in a finite set Φ and the reference measure χ is the counting measure on Φ, i.e. $\chi(\varphi)=1$ for each $\varphi\in\Phi$. If φ, $\psi\in\Omega$ are configurations, then $\varphi=\psi$ (a.s.) means that $\varphi(x)\neq\psi(x)$ holds only on a finite subset of Z^d. The restriction of φ to a subset V of Z^d is denoted by $\varphi(V)$. The cardinality of an arbitrary set ϑ is denoted by $|\vartheta|$, while $|\mu|=\max_{1\leqq i\leqq r}|\mu_i|$ if $\mu=(\mu_1,\mu_2,\ldots,\mu_r)$ is a vector; thus $|x-y|$ is the L_∞-type distance in Z^d, the euclidean distance being denoted by $\|x-y\|$. Further, if V, $W\subset Z^d$ and $x,y\in Z^d$, then $d(x,W)=\min_{y\in W}|x-y|$, $d(V,W)=\min_{x\in V}d(x,W)$, diam $W=\sup_{x,y\in W}|x-y|$, $W_s(x)=\{y\in Z^d\,\big|\,|x-y|\leqq s\}$ is the x-centred cube of radius s, $\bar V=Z^d-V$ is the complement of $V\in Z^d$ and $\partial V=\{x\in V\,|\,d(x,\bar V)=1\}$ defines the boundary of V. We say that V is a connected set in Z^d if any two points $x,y\in V$ are connected by a path $\{x_0,x_1,\ldots,x_n\}\subset V$ in V; i.e. $x=x_0$, $y=x_n$ and $|x_{i+1}-x_i|=1$ for each $i=1,2,\ldots,n-1$.

Consider now an interaction potential $\mathscr{I}$, which is a finite-valued real function $\mathscr{I}=\mathscr{I}(\varphi(V))$ on families $\{\varphi(V)\}$ of finite configurations for each finite $V\subset Z^d$. The formal Hamiltonian $\mathscr{H}$ corresponding to an interaction potential $\mathscr{I}$ is the formal sum

$$\mathscr{H}(\varphi)=\sum\nolimits_{V\subset Z^d}\mathscr{I}(\varphi(V)).$$

The interaction radius R of $\mathscr{H}$ is defined as the minimal number such that $\mathscr{I}(\varphi(V))=0$ whenever diam $V\geqq R$. The formal Hamiltonian can be given also in terms of the one-point interaction energy

$$\mathscr{U}_x(\varphi)=\sum\nolimits_{V:V\ni x}\frac{1}{|V|}\mathscr{I}(\varphi(V))$$

as

$$\mathscr{H}(\varphi)=\sum\nolimits_{x\in Z^d}\mathscr{U}_x(\varphi).$$

In general a one-parameter family $\beta\mathscr{H}$, $\beta\geqq 0$, of Hamiltonians is considered. The conditional Gibbs distribution $P_V(\cdot\,|\psi;\beta\mathscr{H})$ in a finite volume $V\subset Z^d$ with boundary condition $\psi\in\Omega$ is defined by

$$p_V(\varphi|\psi)=\frac{1}{\Theta(V,\beta\mathscr{H},\psi)}\exp\left[-\beta\sum\nolimits_{W:W\cap V\neq 0}\mathscr{I}(\varphi(W))\right]\tag{2.1}$$

if $\varphi(\bar V)=\psi(\bar V)$ and $p_V(\varphi|\psi)=0$ if $\varphi(\bar V)\neq\psi(\bar V)$. The norming factor Θ is the statistical sum, i.e.

(2.2) $$\Theta(V, \beta, \psi) = \sum \exp\left[-\beta \sum_{W:W\cap V\neq 0} \mathscr{I}(\varphi(W))\right];$$

$p_V(\varphi|\psi)$ is the conditional probability of the occurrence of $\varphi\in\Omega$ given that $\varphi(\bar{V})= =\psi(\bar{V})$. It is very convenient to use the notion of relative Hamiltonian $\mathscr{H}(\varphi|\psi)$, if φ, $\psi\in\Omega$ and $\varphi=\psi$ (a.s.), then

(2.3) $$\mathscr{H}(\varphi|\psi) = \sum_{V\subset Z^d}[\mathscr{I}(\varphi(V)) - \mathscr{I}(\psi(V))] = \\ = \sum_{x\in Z^d}[\mathscr{U}_x(\varphi) - \mathscr{U}_x(\psi)].$$

If the interaction radius of $\mathscr{H}$ is finite, then only a finite number of summands do not vanish, thus $\mathscr{H}(\varphi|\psi)$ makes sense. Let us remark that (2.1) can be re-written as

(2.4) $$\frac{p_V(\varphi|\psi)}{p_V(\psi|\psi)} = \exp[-\beta\mathscr{H}(\varphi|\psi)],$$

provided that $\varphi(x)=\psi(x)$ whenever $d(x, \bar{V})$ is less than the interaction radius of $\mathscr{H}$. In this chapter only Hamiltonians of a finite interaction radius will be considered.

Another natural assumption is that the interaction is translation-invariant. This condition, however, can be relaxed without any additional difficulty; we assume only that our Hamiltonians are periodic. More exactly, let $s>0$ and consider a subgroup $\hat{Z}\subset Z^d$ of finite index such that the cardinality of the factor group $Z^d/\hat{Z}$ is less than s. The class $\boldsymbol{H}_s(\hat{Z})$ of Hamiltonians we are dealing with is defined as the set of $\hat{Z}$-periodic Hamiltonians with an interaction radius less than s.

2. Ground states

We say that a one-parameter family P_β, $\beta\to\infty$, of limit Gibbs distributions describes small local distortions of a given configuration $\psi\in\Omega$ if at large values of β the random set $\{x\in Z^d|\varphi(x)\neq\psi(x)\}$ decomposes with probability one into a countable union of finite connected subsets of Z^d, and

(2.5) $$\lim_{\beta\to\infty} \sup_{x\in Z^d} P_\beta\left[\varphi(W_s(x)) \neq \psi(W_s(x))\right] = 0$$

for each $s>0$. This generalizes our intuitive picture that typical configurations in pure phases of the Ising ferromagnet consist of a large "sea" with some rare "islands".

Letting $\beta\to\infty$ in (2.4) we see that typical configurations of the conditional Gibbs distribution at large β are close to configurations of minimal energy. Therefore we may hope that there exist families of limit Gibbs distributions describing small local distortions of some of the stable ground states of our Hamiltonian. To make heuristic considerations clear, we now introduce the concept of ground states and define their stability; this stability property will be referred to as Peierls's condition. Doing this we get into a position to prove the existence of one-parameter families of limit Gibbs distributions describing small local distortions of certain ground states; a more complete characterization of pure phases will be obtained, too. As you shall see however, the proof is rather sophisticated and not at all short.

Definition 2.1. *A configuration $\psi \in \Omega$ is called a ground state of $\mathscr{H}$ if $\mathscr{H}(\varphi|\psi) \geqq 0$ holds for each $\varphi \in \Omega$ with $\varphi = \psi$ (a.s.). We say that ψ is an isolated ground state if $\mathscr{H}(\varphi|\psi) > 0$ whenever $\varphi = \psi$ (a.s.) but $\varphi \neq \psi$.*

Let $\hat{\Omega}$ denote the set of periodic configurations, $g(\mathscr{H})$ is the set of periodic ground states of $\mathscr{H}$. For periodic ground states we have a more intuitive definition. Namely, as the interaction is periodic, the specific energy

$$h(\varphi) = \lim_{s\to\infty} \frac{1}{|W_s(0)|} \sum_{x \in W_s(0)} \mathscr{U}_x(\varphi)$$

is well defined on $\hat{\Omega}$ and also $\underline{h} = \inf_{\varphi \in \Omega} h(\varphi)$ is finite.

Lemma 2.1. *The set $g(\mathscr{H})$ of periodic ground states of $\mathscr{H}$ coincides with*

$$g_0(\mathscr{H}) = \{\psi \in \hat{\Omega} | h(\psi) = \underline{h}\}.$$

Proof. First we show that $g(\mathscr{H}) \subset g_0(\mathscr{H})$. Suppose that $\varphi \in g(\mathscr{H})$, then there exists a $\psi \in \hat{\Omega}$ such that $h(\psi) \leqq \frac{1}{2}(\underline{h} + h(\varphi))$. Let $\varphi^{(s)} = \psi$ on $W_s(0)$ while $\varphi^{(s)} = \varphi$ otherwise, then

$$0 \leqq \mathscr{H}(\varphi^{(s)}|\varphi) = (h(\psi) - h(\varphi)) \cdot |W_s(0)| + O(|\partial W_s(0)|) \leqq$$

$$\leqq \frac{1}{2}(\underline{h} - h(\varphi)) \cdot |W_s(0)| + O(|\partial W_s(0)|)$$

holds as $s \to \infty$, which is possible only if $h(\varphi) = \underline{h}$; i.e. $g(\mathscr{H}) \subset g_0(\mathscr{H})$.

On the other hand, let $\psi \in g_0(\mathscr{H})$, $\psi' = \psi$ (a.s.) and define φ as the periodic continuation of $\psi'(W_s(0))$ to the whole lattice. If s is large enough, then

$$\underline{h}(\varphi) = h(\psi) + \frac{1}{|W_s(0)|} \mathscr{H}(\psi'|\psi),$$

thus $h(\varphi) \geqq h(\psi)$ implies $\mathscr{H}(\psi'|\psi) \geqq 0$, i.e. $\psi \in g(\mathscr{H})$. Q.e.d.

Intuitively it is clear that only an isolated ground state can be associated with a low temperature pure phase. Peierls's condition requires that the periodic ground states are uniformly isolated in a certain sense; i.e. they are stable. This condition plays a central role in the forthcoming investigations.

Definition 2.2. *Suppose that we are given a positive integer s and a finite set $g=\{\psi_1, \psi_2, \ldots, \psi_r\}$ of periodic configurations. Then the boundary $\partial\varphi$ of a configuration $\varphi\in\Omega$ is defined as*

$$\partial\varphi = \bigcup_{x\in Z^d}\{W_s(x)|\varphi(W_s(x)) \neq \psi(W_s(x)) \text{ for each } \psi\in g\}.$$

The boundary is not necessarily a finite set but it is so if $\varphi=\psi$ (a.s.) for some $\psi\in g$. The definition of $\partial\varphi$ depends on s and g, but the dependence on s is not too important. Indeed, if $\partial'\varphi$ denotes the boundary defined by means of s' and g, where $s'>s$, then

$$\partial\varphi \subset \partial'\varphi \subset \bigcup_{x\in\partial\varphi} W_{s'}(x),$$

so that

$$|\partial\varphi| \leqq |\partial'\varphi| \leqq [4s'+1]^d\,|\partial\varphi|. \tag{2.6}$$

Peierls's condition is formulated in the following set up. We are given a Hamiltonian $\mathscr{H}_0$ with a finite set $\{\psi_1, \psi_2, \ldots, \psi_r\}=g(\mathscr{H}_0)$ of periodic ground states. The boundary $\partial\varphi$ of configurations is defined by means of $g=\{\psi_1, \psi_2, \ldots, \psi_r\}$, the constant s appearing in the definition of $\partial\varphi$ will be specified later. Since $h(\psi)=\underline{h}_0$ on $g(\mathscr{H}_0)$, one would expect that $\mathscr{H}_0(\varphi|\psi)$ is approximately proportional to $|\partial\varphi|$ if $\psi\in g(\mathscr{H}_0)$. The existence of a uniform upper bound

$$\mathscr{H}_0(\varphi|\psi) \leqq C|\partial\varphi| \quad \text{if} \quad \psi\in g(\mathscr{H}_0), \quad \varphi\in\Omega$$

follows immediately; however, we need a lower bound.

Definition 2.3. *Our Hamiltonian $\mathscr{H}_0$ satisfies Peierls's condition if there exists a $\varrho>0$ such that $\mathscr{H}_0(\varphi|\psi)\geqq\varrho|\partial\varphi|$ for each $\psi\in g(\mathscr{H}_0)$ and $\varphi=\psi$ (a.s.).*

Bearing in mind (2.6) we see that the validity of Peierls's condition does not depend on the choice of s; ϱ remains positive as s gets larger. Since $|\partial\varphi|>0$ if $\varphi\notin g(\mathscr{H}_0)$, Peierls's condition implies that periodic ground states of $\mathscr{H}_0$ are isolated.

Peierls's condition has been verified in many cases. The following assertion seems to be true, but it has not been proven yet.

Conjecture: *$|g(\mathscr{H}_0)|<+\infty$ implies Peierls's condition.*

3. Ground states of the perturbed Hamiltonian

Let $\mathscr{H}_0$ denote a periodic Hamiltonian with a finite radius of interaction and suppose that the set $g(\mathscr{H}_0)$ of periodic ground states of $\mathscr{H}_0$ is a nonempty finite set. From now on we fix a subgroup $\hat{Z}\subset Z^d$ of finite index and a positive integer s in such a way that each $\psi\in g(\mathscr{H}_0)$ and any other Hamiltonian $\mathscr{H}$ we are to consider later belong to $\boldsymbol{H}_s(\hat{Z})$; for the definition of $\boldsymbol{H}_s(\hat{Z})$ see the end of Section 1. The boundary $\partial\varphi$ of configurations $\varphi\in\Omega$ is defined in terms of $g(\mathscr{H}_0)$ and of s above; let $g(\mathscr{H}_0)=$

$=\{\psi_1, \dots, \psi_r\}$. Our main assumption, i.e. Peierls's condition, ensures that

$$\mathscr{H}_0(\varphi|\psi) \geqq \varrho\,|\partial\varphi| \quad \text{if} \quad \varphi\in\Omega, \quad \psi\in g(\mathscr{H}_0), \tag{2.7}$$

where ϱ is a positive constant.

The following lemma characterizes the set of ground states for small perturbations $\mathscr{H}=\mathscr{H}_0+\tilde{\mathscr{H}}$ of $\mathscr{H}_0$. The magnitude of a perturbation is given by

$$\|\tilde{\mathscr{H}}\| = \sup_{x\in Z^d}\sup_{\varphi\in\Omega}|\tilde{\mathscr{U}}_x(\varphi)|, \tag{2.8}$$

where $\tilde{\mathscr{U}}$ is the one-point interaction energy corresponding to $\tilde{\mathscr{H}}$. Let us remark that $\boldsymbol{H}_s(\hat{Z})$ is a finite-dimensional Banach space with norm $\|\cdot\|$. Just as earlier, $h(\varphi)$, $h_0(\varphi)$ and $\tilde{h}(\varphi)$ denote the specific energy of $\varphi\in\hat{\Omega}$ with respect to $\mathscr{H}$, $\mathscr{H}_0$, and $\tilde{\mathscr{H}}$, respectively; $\underline{\tilde{h}}=\min_{1\leqq q\leqq r}\tilde{h}(\psi_q)$.

Lemma 2.2. *Suppose that $\tilde{\mathscr{H}}\in\boldsymbol{H}_s(\hat{Z})$ and $\|\tilde{\mathscr{H}}\|<e^{-a}\varrho$. Then $g(\mathscr{H}_0+\tilde{\mathscr{H}})\subset g(\mathscr{H}_0)$ and*

$$g(\mathscr{H}_0+\tilde{\mathscr{H}}) = \{\psi\in g(\mathscr{H}_0)|\tilde{h}(\psi)=\underline{\tilde{h}}\}$$

where $a\geqq d\ln(2s+1)$ is a universal constant to be specified in Lemma 2.7.

Proof. Let $\varphi\in\hat{\Omega}$ and introduce

$$\mathrm{Int}_q(\varphi) = \{x\in Z^d-\partial\varphi|\varphi(W_s(x)) = \psi_q(W_s(x))\},$$

then we have

$$\Big|\sum_{x\in W_L(0)}\tilde{\mathscr{U}}_x(\varphi)-\sum_{q=1}^{r}|\mathrm{Int}_q(\varphi)\cap W_L(0)|\tilde{h}(\varphi)\Big| \leqq |(2s+1)^d-1|\,\|\tilde{\mathscr{H}}\|\,|\partial\varphi|. \tag{2.9}$$

Letting $L\to\infty$ we obtain

$$\Big|\tilde{h}(\varphi)-\sum_{q=1}^{r}\Pi_q(\varphi)\tilde{h}(\psi_q)\Big| \leqq [(2s+1)^d-1]\,\|\tilde{\mathscr{H}}\|\,\Pi_0(\varphi), \tag{2.10}$$

where

$$\Pi_0(\varphi) = \lim_{L\to\infty}\frac{|\partial\varphi\cap W_L(0)|}{|W_L(0)|} \tag{2.11}$$

and

$$\Pi_q(\varphi) = \lim_{L\to\infty}\frac{|\mathrm{Int}_q(\varphi)\cap W_L(0)|}{|W_L(0)|} \tag{2.12}$$

if $1\leqq q\leqq r$. On the other hand, (2.7) turns into

$$h_0(\varphi) \geqq h_0(\psi)+\varrho\,\Pi_0(\varphi) \tag{2.13}$$

whenever $\psi\in g(\mathscr{H}_0)$; thus

$$\sum_{q=0}^{r}\Pi_q(\varphi) = 1, \quad \Pi_q(\varphi)\geqq 0 \tag{2.14}$$

and $h(\psi)=h_0(\psi)+\tilde{h}(\psi)$ imply

$$h(\varphi) = h_0(\varphi)+\tilde{h}(\varphi) \geqq$$
$$\geqq h_0(\varphi)+\Pi_0(\varphi)\left[\varrho-[(2s+1)^d-1]\|\tilde{\mathscr{H}}\|\right]+\textstyle\sum_{q=1}^{r}\Pi_q(\varphi)\tilde{h}(\psi_q)=$$
$$= h(\psi)-\textstyle\sum_{q=0}^{r}\Pi_q(\varphi)\tilde{h}(\psi)+\sum_{q=1}^{r}\Pi_q(\varphi)\tilde{h}(\psi_q)+$$
$$+\Pi_0(\varphi)\left[\varrho-[(2s+1)^d-1]\|\tilde{\mathscr{H}}\|\right].$$

Since $|\tilde{h}(\psi)|\leqq\|\tilde{\mathscr{H}}\|$, we have that

$$h(\varphi)\geqq h(\psi)+\textstyle\sum_{q=1}^{r}\Pi_q(\varphi)[\tilde{h}(\psi_q)-\tilde{h}(\psi)]+\Pi_0(\varphi)[\varrho-(2s+1)^d\|\tilde{\mathscr{H}}\|] \tag{2.15}$$

holds for each $\varphi\in\hat{\Omega}$ and $\psi\in g(\mathscr{H}_0)$. Choosing $\psi\in g(\mathscr{H}_0)$ in such a way that $\tilde{h}(\psi)=\tilde{h}$, we obtain $\tilde{h}(\varphi)\geqq\tilde{h}$ for each $\varphi\in\hat{\Omega}$, furthermore $\tilde{h}(\varphi)>\tilde{h}$ follows directly from (2.15) if $\varphi\notin g(\mathscr{H}_0)$. Thus Lemma 2.1 implies the statement. Q.e.d.

Now we are in a position to specify the perturbation $\mathscr{H}=\mathscr{H}_0+\tilde{\mathscr{H}}$ of the basic Hamiltonian $\mathscr{H}_0$ we shall consider later. Namely, let $\mathscr{H}_\mu=\mathscr{H}_0+\sum_{i=1}^{r-1}\mu_i\mathscr{H}_i$, where $\mathscr{H}_i\in\boldsymbol{H}_s(\hat{Z})$ for each $i=1, 2, \ldots, r-1$, and suppose that the external fields $\mathscr{H}_i$ are linearly independent in the following sense.

Definition 2.4. *Let* $t_\mu=(t_\mu(\psi_1), \ldots, t_\mu(\psi_r))$ *where* $t_\mu(\psi_q)=h_\mu(\psi_q)-\min_{1\leqq m\leqq r}h_\mu(\psi_m)$, *and* h_μ *denotes specific energy with respect to* $\mathscr{H}_\mu$. *We say that the perturbation* $\mathscr{H}_\mu=\mathscr{H}_0+\sum_{i=1}^{r-1}\mu_i\mathscr{H}_i$ *completely splits the degeneracy of the ground state of* $\mathscr{H}_0$ *if* t_μ *maps the space of parameters onto the entire boundary* $0_r=\{b|b=(b_1, \ldots, b_r), \min_{1\leqq q\leqq r}b_q=0\}$ *of the r-dimensional positive octant.*

This condition means that any nonempty subset of $g(\mathscr{H}_0)$ appears as the set of ground states of $\mathscr{H}_\mu$ when μ runs over a small neighbourhood of 0 in the parameter space (cf. Lemma 2.2). As we shall see later, Peierls's condition implies that for large β and small μ, the topological structure of the set $\mathscr{G}(\beta\mathscr{H}_\mu)$ of indecomposable and periodic limit Gibbs distributions is the same as that of 0_r, i.e. we have a regular phase diagram.

4. Phase transitions in the two-dimensional Ising ferromagnet

Now we stop for a while in building up the general theory and go back to the basic example which has been the starting point of the development of the whole theory treated in this chapter. This famous example is the two-dimensional Ising ferromagnet, that is the lattice model with Hamiltonian

$$\mathscr{H}_0=-\mathscr{J}\textstyle\sum_{\|x-y\|=1}\varphi(x)\varphi(y),\quad \varphi(x)=\pm1,\quad x, y\in Z^2,$$

where the sum is over pairs $x, y \in Z^2$ with $\|x-y\|=1$; $\mathscr{I}>0$ in the ferromagnetic case, we may and do assume that $\mathscr{I}=1$. It is easy to check that $\mathscr{g}(\mathscr{H}_0)=\{\psi^+, \psi^-\}$ with $\psi^+=\{\varphi(x)\equiv 1\}$, $\psi^-=\{\varphi(x)\equiv -1\}$, and that $\mathscr{H}_0$ satisfies Peierls's condition.

Theorem 2.1. (Peierls [105]). *There exists a $\beta>0$ such that for $\beta>\beta_0$ there exist at least two translation-invariant limit Gibbs distributions. As functions of β, these Gibbs distributions describe small local distortions of the ground states ψ^+ and ψ^-, respectively.*

Proof. Fix a volume $V\subset Z^2$ and the boundary condition $\bar{\varphi}(\bar{V})$ outside V. Then the conditional Gibbs distribution in V can be written as

$$P_\beta(\varphi(V)|\bar{\varphi}(\bar{V})) = \frac{1}{\Theta(V; \beta; \bar{\varphi}(\bar{V}))} \exp\{\beta \sum_{\substack{\|x-y\|=1\\ \{x,y\}\cap V\neq\emptyset}} \varphi(x)\varphi(y)\}$$

where Θ is the statistical sum, $\beta>0$ is the inverse temperature and V is chosen as a square $V=W_L(0)$. We consider the boundary condition $\bar{\varphi}(\bar{V})\equiv +1$; the corresponding conditional Gibbs distribution is denoted by P_V^+.

In this case the boundary $\partial(\varphi(V))$ of a finite configuration $\varphi(V)$ can be defined in a simpler but slightly different way than in the general situation. Let E_x denote the x-centred closed unit square in R^2, the sides of E_x are parallel to the coordinate axes. Then $\partial(\varphi(V))$ is simply the boundary of the union of such E_x that $x\in V$ and $\varphi(x)=-1$. Contours of $\varphi(V)$ are those closed trajectories in $\partial(\varphi(V))$ which do not intersect themselves. Each boundary $\partial(\varphi(V))$ decomposes into contours in a unique way, and the correspondence $\varphi(V)\to\partial(\varphi(V))$ between configurations and their boundaries is one-to-one. Indeed, if we know $\partial(\varphi(V))$, then proceeding from the boundary of V inside, we put $+1$ everywhere until we cross the first contour; after crossing it we put -1 and so on; this way the entire configuration can be reconstructed.

The proof of the theorem is based on the following combinatorial result.

Peierls's inequality (*see* [105]). *Let γ denote a fixed contour, then*

$$P_V^+(\gamma\subset\partial(\varphi(|V|)))\leqq e^{-2\beta|\gamma|}$$

where $|\gamma|$ is the length of γ in the usual sense.

Proof. If $\varphi=\psi^+$ outside V, then

$$\mathscr{H}_V(\varphi) = \mathscr{H}(\varphi(V))+\mathscr{H}(\varphi(V)|\psi^+(\bar{V})) =$$
$$= -2|V|+2|\partial(\varphi(V))|$$

where $|\partial(\varphi(V))|$ denotes the joint length of contours in $\partial(\varphi(V))$. Namely,

$$\mathscr{H}_V(\varphi) = -\sum_{\{x,y\}:\{x,y\}\cap V\neq\emptyset} \varphi(x)\varphi(y) = -\sum{}^+ - \sum{}^-,$$

where the summation in $\sum^+$ and in $\sum^-$ is over such pairs that $\varphi(x)=\varphi(y)$, and $\varphi(x)=-\varphi(y)$, respectively. In view of the definition of $\partial(\varphi(V))$ we have $-\sum^-=$ $=|\partial(\varphi(V))|$ because the side of an E_x separating a $+1$ and -1 belongs to $\partial(\varphi(V))$. On the other hand, the total number of pairs $\{x,y\}$ such that $\{x,y\}\cap V\neq\emptyset$, $\|x-y\|=1$ is just $2|V|$, thus $\sum^+=2|V|+\sum^-$, i.e. $\mathscr{H}_V(\varphi)=-2|V|-2\sum^-=-2|V|+$ $+2|\partial(\varphi(V))|$. Further

$$P_V^+(\gamma\subset\partial(\varphi(V)))=\frac{\sum_{\varphi(V):\gamma\subset\partial(\varphi(V))}\exp(-\beta\mathscr{H}_V(\varphi))}{\sum_{\varphi(V)}\exp(-2\beta\mathscr{H}_V(\varphi))}=$$

$$=\frac{\sum_{\varphi(V):\gamma\subset\partial(\varphi(V))}\exp(-2\beta|\partial(\varphi(V))|)}{\sum_{\varphi(V)}\exp(-2\beta|\partial(\varphi(V))|)}$$

follows from the above relation and from the definition of P_V^+. Let Φ_γ denote the set of such configurations that $\gamma\subset\partial(\varphi(V))$, Φ_γ^- is the set of configurations for which $\partial(\varphi(V))\cap\gamma=\emptyset$. Now we define a one-to-one correspondence χ_γ between Φ_γ and Φ_γ^- by changing the sign of each $\varphi(x)$ if x is surrounded by γ. It is plain that χ_γ removes γ from the boundary of $\varphi(V)\in\Phi_\gamma$; i.e.

$$\partial\varphi(V)=\partial(\chi_\gamma\varphi(V))\cup\gamma$$

and

$$|\partial\varphi(V)|=|\partial(\chi_\gamma\varphi(V))|+|\gamma|.$$

Therefore

$$P_V^+(\gamma\subset\partial(\varphi(V)))=\frac{\sum_{\varphi(V)\in\Phi_\gamma}\exp(-2\beta|\partial(\varphi(V))|)}{\sum_{\varphi(V)}\exp(-2\beta|\partial(\varphi(V))|)}=$$

$$=e^{-2\beta|\gamma|}\frac{\sum_{\varphi(V)\in\Phi_\gamma^-}\exp(-2\beta|\partial(\varphi(V))|)}{\sum_{\varphi(V)}\exp(-2\beta|\partial(\varphi(V))|)}\leqq e^{-2\beta|\gamma|}$$

which proves Peierls's inequality. Q.e.d.

We first prove two corollaries to this inequality.

Corollary 1. $\lim_{\beta\to\infty}P_V^+(\varphi(0)=-1)=0$ *uniformly in* V.

Proof. If $\varphi(0)=-1$, then the origin 0 of Z^2 is surrounded by at least one contour γ, we indicate this relation as $0\in\operatorname{Int}\gamma$. Therefore Peierls's inequality implies immediately that

$$P_V^+(\varphi(0)=-1)\leqq\sum_{\gamma:0\in\operatorname{Int}\gamma}e^{-2\beta|\gamma|}.$$

Observe that $|\gamma|\geqq 4$ if γ is a contour, thus

$$P_V^+(\varphi(0)=-1)\leqq\sum_{n=4}^{\infty}C_n e^{-2\beta n},$$

where C_n is the number of such contours γ that $|\gamma|=n$ and $0\in\operatorname{Int}\gamma$. To estimate C_n, let $z(\gamma)$ denote the first point of γ on the positive half of the horizontal coordinate axis; if $0\in\operatorname{Int}\gamma$ then $z(\gamma)$ is well defined. In case of $|\gamma|=n$, $z(\gamma)$ can be chosen in not more than n different ways, and the number of contours of length n and passing through a fixed point z is certainly not more than 3^n. Indeed, the possible ways for the continuation of a contour at one of its vertices is at most 3. Therefore $C_n \leqq n3^n$, thus

$$P_V^+(\varphi(0)=-1) \leqq \sum_{n=4}^{\infty} n3^n e^{-2\beta n},$$

where the series on the right-hand side converges if $\beta > \frac{1}{2}\ln 3$, and its convergence is uniform in β if β is bounded away from this value. The above bound is independent of V, so the Corollary follows directly by the dominated convergence theorem. Q.e.d.

Corollary 2. *If β is large enough, then there exists a constant $c=c(\beta)<0$ such that $\lim_{\beta\to\infty} c(\beta)=0$ and*

$$\lim_{V\to\infty} P_V^+\left[|\gamma| > c \ln |V| \text{ for some } \gamma \in \partial(\varphi(V))\right] = 0.$$

Proof. Peierls's inequality implies in a similar way as above that

$$P_V^+ \leqq |V| \sum_{n > c\ln|V|} 3^n e^{-2\beta n} \leqq$$

$$\leqq |V| \frac{\exp(\ln 3 - 2\beta) c \ln V]}{1-\exp[\ln 3 - 2\beta]}. \quad \text{Q.e.d.}$$

Now we are in a position to prove the theorem. Consider the sequence P_V^+ of conditional Gibbs distributions in $V=W_L(0)$ with boundary condition $\varphi \equiv +1$. Since Ω is a compact space, this sequence certainly has at least one limit point P^+ as $L\to\infty$. Q.e.d.

In view of Corollary 1 we have

$$p^+(\varphi(0)=-1) < \frac{1}{2}$$

if β is large enough.

Taking into account the symmetry $+1 \to -1$ of $\mathscr{H}_0$, we obtain in the same way that there exists such a limit Gibbs state P^- corresponding to the boundary condition $\bar{\varphi} \equiv -1$ that $P^-(\varphi(0)=+1) < \frac{1}{2}$ if β is large enough, i.e.

$$P^-(\varphi(0)=-1) > \frac{1}{2}.$$

This means that $P^+ \neq P^-$, i.e. at large values of β there exist at least two limit Gibbs states for the Hamiltonian $\beta\mathscr{H}_0$. For P^+ a typical configuration consists of a "sea" of $+1$ with some small "islands" of -1; for P^- the situation is the opposite. It is easy to deduce from Corollary 1 that P^+ and P^- describe small distortions of the ground states ψ^+ and ψ^-, respectively. Q.e.d.

5. The Main Theorem and its consequences

Now we return to the general problem of phase transitions in classical lattice systems as posed in Section 3. Suppose we are given a periodic Hamiltonian $\mathscr{H}_0$ with a finite interaction radius, and the set $g(\mathscr{H}_0)=\{\psi_1, \psi_2, \ldots, \psi_r\}$ of periodic ground states of $\mathscr{H}_0$ is finite and $\mathscr{H}_0$ satisfies Peierls's stability condition. We consider a small perturbation $\mathscr{H}_\mu=\mathscr{H}_0+\mu_1\mathscr{H}_1+\ldots+\mu_{r-1}\mathscr{H}_{r-1}$, $\mu=(\mu_1, \mu_2, \ldots, \mu_{r-1})$ of the basic Hamiltonian $\mathscr{H}_0$, where the external fields $\mathscr{H}_i$ $(i=1, 2, \ldots, r-1)$ are arbitrary periodic Hamiltonians of a finite radius of interaction such that the perturbation $\mathscr{H}_\mu$ completely splits the degeneracy of the ground state of $\mathscr{H}_0$. Now we fix an integer $s>0$ and a subgroup $\hat{Z}\subset Z^d$ such that each ground state $\psi_q\in g(\mathscr{H}_0)$ is $\hat{Z}$-periodic and $\mathscr{H}_i\in \boldsymbol{H}_s(\hat{Z})$ for each $i=0, 1, \ldots, r-1$; the boundary $\partial\varphi$ of configurations will be defined by means of s and $g(\mathscr{H}_0)$.

Assume, at least for large values of β, the existence of the weak limit

$$P^{(q)}_{\beta,\mu} = \lim_{V\to\infty} P_V[\cdot\,|\psi_q, \beta\mathscr{H}_\mu]$$

of conditional Gibbs distributions, and that the family $P^{(q)}_{\beta,\mu}$, $\beta\to\infty$, of limit Gibbs distributions describes small local distortions of ψ_q. Then we say that $p^{(q)}$ is a *pure phase associated with* ψ_q. Our construction of pure phases (see Propositions 2.6) implies that they are $\hat{Z}$-periodic ergodic random fields with an exponential decay of correlations, so that a pure phase in the above sense is really an indecomposable and periodic limit Gibbs distribution. It is plain that pure phases associated with different ground states are different.,

From now on we assume that the dimension d of our system is at least two. This condition is needed to ensure that a large distortion of a ground state needs a large amount of energy, i.e. $\mathscr{H}(\varphi|\psi_q)\to\infty$ if $\{x\in Z^d|\varphi(x)\neq\psi_q(x)\}\to\infty$.

Main Theorem A. *There exist positive constants β_0 and ε_0 depending only on $\mathscr{H}_0$, $\mathscr{H}_1, \ldots, \mathscr{H}_{r-1}$ and on $d\geqq 2$ such that, for $\beta\geqq\beta_0$ and $\mu\in\mathscr{U}_0=\{\mu||\mu|<\varepsilon_0\}$, we have the following picture.*

1. *There exists a point $\bar{\mu}(\beta)\in\mathscr{U}_0$ such that each limit Gibbs distribution $P^{(q)}_{\beta,\bar{\mu}(\beta)}$, $1\leqq q\leqq r$, is a different pure phase.*

2. *There exist such open trajectories $\gamma_1(\beta), \gamma_2(\beta), \ldots, \gamma_r(\beta)\subset\mathscr{U}_0$ starting from $\bar{\mu}(\beta)$ that $P^{(1)}_{\beta,\mu}, \ldots, P^{(q-1)}_{\beta,\mu}, P^{(q+1)}_{\beta,\mu} \ldots P^{(r)}_{\beta,u}$ are different pure phases if $\mu\in\gamma_q(\beta)$.*

3. *There exist such two-dimensional open surfaces $\gamma_{ij}(\beta)\subset\mathscr{U}_0$, $1\leqq i\leqq j\leqq r$, that the boundary of $\gamma_{ij}(\beta)$ consists of $\gamma_i(\beta), \gamma_j(\beta), \bar{\mu}(\beta)$, and the limit Gibbs states $P^{(q)}_{\beta,\mu}$ are different pure phases for $\mu\in\gamma_{ij}(\beta)$ whenever $q\neq i$ and $q\neq j$.*

There exist such disjoint open subsets $\gamma^{(1)}(\beta), \ldots, \gamma^{(r)}(\beta)$ of $\mathscr{U}_0$, bounded by the $(r-2)$-dimensional surfaces of coexistence of two pure phases (statement $r-1$), that their closures jointly cover $\mathscr{U}_0$, and $P^{(q)}_{\beta,\mu}$ is a pure phase if $\mu\in\gamma^{(q)}(\beta)$.

The following statement is a somewhat stronger form of Main Theorem A.

Main Theorem B. *There exist such positive constants* β_0 *and* ε_0 *depending only on* $\mathcal{H}_0, \mathcal{H}_1, \ldots, \mathcal{H}_{r-1}$ *and on* $d \geqq 2$ *that for each* $\beta \geqq \beta_0$ *a homomorphism* $I_\beta = I_\beta(\mu)$ *of* $\mathcal{U}_0 = \{\mu \,|\, |\mu| < \varepsilon_0\}$ *into* 0_r *is defined with the following properties. The image* $I_\beta \mathcal{U}_0$ *of* $\mathcal{U}_0$ *contains an entire neighbourhood of* 0 *in* 0_r*; further for every* q *such that* $b_q = 0$ *in* $I_\beta(\mu) = (b_1, b_2, \ldots, b_r)$*, the limit Gibbs distributions* $P^{(q)}_{\beta,\mu}$ *exist and are different pure phases.*

Let us remark that if $\mathcal{H}_0$ possesses some symmetry properties, then the mapping I_β describing the phase diagram has the very same symmetry. To formulate this useful consequence in a more exact way, let G denote a subgroup of permutations of the individual state space Φ; the action of the corresponding transformation group induced by G in the configuration space Ω and in the space of Hamiltonians will be denoted in the same way. Suppose now that $\mathcal{H}_0$ is invariant under G and G maps the family $\mathcal{H}_\mu$ onto itself.

It is clear that G is a permutation group also in the set $g(\mathcal{H}_0)$ of ground states, thus $\psi_{gq} = g\psi_q$, $g \in G$, defines a permutation of the index set $\{1, 2, \ldots, r\}$, while $g\mathcal{H}_\mu = \mathcal{H}_{g\mu}$ defines the action of G in the parameter space. We say that our mapping I_β is G-invariant if $I_\beta(g\mu) = gI_\beta(\mu)$ holds for each μ and $g \in G$; here gI_β indicates that the coordinates $b_1, b_2, \ldots, b_r$ of $I_\beta(\beta)$ have been rearranged by the permutation g of $\{1, 2, \ldots, r\}$. Having a look at the proof of the Main Theorem one can check easily that G-invariance of $\mathcal{H}_0$ and of the family $\mathcal{H}_\mu$ implies that of I_β in the above sense. This simple fact will be very useful in deducing some consequences of the basic theorem, as in the following examples.

1. *Small perturbations of the Ising ferromagnet.*

In this case $\Phi = \{+1, -1\}$, $d \geqq 2$,

$$\mathcal{H}_0(\varphi) = -\sum\nolimits_{x \in Z^d} \sum\nolimits_{\|x-y\|=1} \varphi(x)\,\varphi(y),$$

$$\mathcal{M}(\varphi) = \sum\nolimits_{x \in Z^d} \varphi(x)$$

and $\mathcal{H} = \mathcal{H}_0 + \varepsilon\mathcal{H}_1 + h\mathcal{M}$, where $\mathcal{H}_1$ is such a Hamiltonian that $h_1(\psi^+) = h_1(\psi^-)$ for the ground states $\psi^+ \equiv +1$ and $\psi^- \equiv -1$ of $\mathcal{H}_0$; h_1 denotes specific energy with respect to $\mathcal{H}_1$. Then $g(\mathcal{H}_0 + \varepsilon\mathcal{H}_1) = \{\psi^+, \psi^-\}$ if ε is small enough, further $\mathcal{H}_0 + \varepsilon\mathcal{H}_1$ satisfies Peierls's condition, and $|g(\mathcal{H})| = 1$ if $h \neq 0$; i.e. the degeneracy of the ground state of $\mathcal{H}_0 + \varepsilon\mathcal{H}_1$ if completely split by the perturbation $\mathcal{H} = \mathcal{H}_0 + \varepsilon\mathcal{H}_1 + h\mathcal{M}$. Therefore we have a phase transition without any breakdown of symmetry:

Theorem 2.2. (see [35]). *There exist positive constants* β_0, ε_0 *and a function* $h = h(\beta, \varepsilon)$ *such that for* $\beta > \beta_0$, $|\varepsilon| < \varepsilon_0$ *and* $h = h(\beta, \varepsilon)$ *there exist at least two pure phases for* $\beta\mathcal{H}$.

2. *Ising ferromagnet with several spin values.*

Let $\Phi=\{-r, -r+1, \ldots, 0, 1, \ldots, r\}$, $d \geqq 2$ and

$$\mathcal{H}_0 = -\sum_{x\in Z^d}\sum_{\|x-y\|=1}\delta(\varphi(x), \varphi(y)),$$

where $\delta(p, q)=1$ if $p=q$, $\delta(p, q)=0$ if $p\neq q$, $p, q\in\Phi$. Then $g(\mathcal{H}_0)=\{\psi_q | q\in\Phi\}$ with $\psi_q\equiv q$. It is easy to check that Peierls's condition also holds in this case.

Theorem 2.3. *There exists a $\beta_0>0$ such that each $P_\beta^{(q)}$, $q\in\Phi$, is a pure phase of $\beta\mathcal{H}_0$, and they are all different.*

Proof. In order to split up the degeneration of $g(\mathcal{H}_0)$ we introduce the external fields

$$\mathcal{H}_q = \sum_{x\in Z^d}\delta(\varphi(x), q)\operatorname{sign}\varphi(x),\ q\in\Phi-\{0\}.$$

It is easy to check that the perturbation

$$\mathcal{H}_\mu = \mathcal{H}_0 + \sum_{i\neq 0}\mu_i\mathcal{H}_i$$

really splits the degeneracy of $g(\mathcal{H}_0)$, consequently for large values of β there exists a point $\bar\mu=\bar\mu(\beta)$ of coexistence of $2r+1$ different pure phases. Since any permutation of $\Phi-\{0\}$ is a symmetry of $\mathcal{H}_0$ and it maps the family $\mathcal{H}_\mu$ onto itself, the coordinates of $\bar\mu$ are necessarily the same. On the other hand, the reflection $q\to -q$ of Φ induces the mapping $\mu\to-\mu$ in the space of parameters. Thus $\bar\mu=0$. Q.e.d.

3. *Ising antiferromagnet.*

Let $\Phi=\{-1, +1\}$, $d\geqq 2$ and

$$\mathcal{H}_0 = \sum_{x\in Z^d}\sum_{y:\|y-x\|=1}\varphi(x)\varphi(y) + h\sum_{x\in Z^d}\varphi(x).$$

Since

$$\mathcal{H}_0 = \sum_{x\in Z^d}\sum_{y:\|y-x\|=1}\left[\varphi(x)\varphi(y)+\frac{h}{4d}(\varphi(x)+\varphi(y))\right] =$$

$$= \frac{1}{2}\sum_{x\in Z^d}\sum_{y:\|y-x\|=1}\left[(\varphi(x)+\varphi(y))^2+\frac{h}{2d}[\varphi(x)+\varphi(y)]-2\right]$$

it is clear that for $|h|<4d$ any periodic ground state of $\mathcal{H}_0$ is one of the two chess-board-type configurations ψ_1, ψ_2 characterized by $\psi_i(x)=-\psi_i(y)$ if $\|x-y\|=1$. Further, Peierls's condition holds with a constant proportional to $4d-|h|$.

Theorem 2.4. (see [13]–[15]). *Let $|h|<4d$. Then there exists a $\beta_0=\beta_0(h)$ such that for $\beta>\beta_0$ both ground states of $\mathcal{H}_0$ are associated with pure phases describing small distortions of the corresponding ground state if $\beta\to\infty$.*

Proof. It is easy to split the degeneracy of $g(\mathcal{H}_0)$ by a suitably chosen perturbation $\mathcal{H}_\mu=\mathcal{H}_0+\mu\mathcal{H}_1$. Thus at $\mu=0$ at least one of ψ_1 and ψ_2 is associated with a pure phase P_β. Since the ψ_i are periodic with period 2, so is P_β (see Proposition 2.6 later). Thus the translate of P_β defines the other pure phase. Q.e.d.

6. Contours

A limit Gibbs distribution associated with a ground state $\psi\in\Omega$ is constructed as the weak limit of conditional Gibbs distributions in finite volumes V with boundary condition $\varphi(\bar{V})=\psi(\bar{V})$ as $V\to\infty$. Since for fixed ψ, $H(\varphi|\psi)$ reduces to being a function only of the pair $(\partial\varphi, \varphi(\partial\varphi))$, the boundary of a configuration plays an important role in the study of such distributions. Namely the components of the boundary separating domains occupied by different ground states are of fundamental importance. Such components are called contours, and they are a straightforward generalization of the contour in two-dimensional Ising ferromagnets. Here and in what follows, general notations introduced in the preceding sections will be used without any reference; the boundary of a configuration is defined by means of $g(\mathscr{H}_0)=\{\psi_1, ..., \psi_r\}$ and of the positive integer s specifying the class $\boldsymbol{H}_s(\hat{Z})$ of Hamiltonians.

In the definition of contours some topological properties of Z^d, $d\geqq 2$ are exploited. Nevertheless, our methods can be extended to other kinds of lattices, too.

We say that two subsets M_1 and M_2 of Z^d are distant if $d(M_1, M_2)>1$. A subset M of a set $W\subset Z^d$ is called a component of W if M is a maximal connected subset of W; i.e. M is connected and $M\subset M'\subset W$, $M\neq M'$, imply that M' cannot be connected. Let us remark that any subset of Z^d is the union of its components, this decomposition is unique, and different components are distant.

Definition 2.5. *Let M denote a finite connected subset of Z^d, and let $\varphi\in\Omega$ be a configuration. Then the couple $\Gamma=(M, \varphi(M))$ is called a contour of φ if M is a component of the boundary $\partial\varphi$ of φ. A couple $\Gamma=(M, \varphi(M))$ of this type is called a contour if there exists at least one configuration $\varphi\in\Omega$ such that Γ is a contour of φ.*

If $\Gamma=(M, \varphi(M))$ is a contour, then M is called the support of Γ, this relation will be indicated as $M=\text{supp}\,\Gamma$. Of course, the configuration on the support of a contour cannot be arbitrary, because it necessarily coincides with one of the ground states near the boundary of the support. Suppose that $\Gamma=(M, \varphi(M))$ is a contour and consider the components A_α of $\bar{M}=Z^d-M$. We show that to each component A_α there corresponds a unique ground state $\psi_{q(A_\alpha)}\in g(\mathscr{H}_0)$. Since A_α can contain other contours, we define the number $q(A_\alpha)$ in terms of $\partial A_\alpha=\{x\in A_\alpha | d(x, M)=1\}$. Observe first that the constant s in the definition of the boundary of configurations is so large that the ground states $\psi_q, \psi_m\in g(\mathscr{H}_0)$ necessarily coincide whenever $\psi_q(W_s(x))=\psi_m(W_s(y))$ holds for at least one pair $x, y\in Z^d$ with $|x-y|=1$. On the other hand, $\varphi(W_s(x))=\psi(W_s(x))$ with $\psi\in g(\mathscr{H}_0)$ if $x\in\partial A_\alpha$ and Γ is a contour of $\varphi\in\Omega$. Since the L_∞-type norm $|\cdot|$ is used in the definition of connectedness, ∂A_α is a connected set and there exists a unique ground state $\psi_q\in g(\mathscr{H}_0)$ such that $\varphi(W_s(x))=\psi_q(W_s(x))$ holds whenever $x\in\partial A_\alpha$ and Γ is a contour of $\varphi\in\Omega$. This means that given Γ, $q(A_\alpha)$ is uniquely defined by $\psi_q=\psi_{q(A_\alpha)}$, where ψ_q is the ground state chosen above. If

$d \geqq 2$, as we do assume, then exactly one of the components A_α of Z^d — supp Γ is infinite, this one will be denoted by Ext Γ, while $\mathrm{Int}_m \Gamma$ is the union of components A_α of Z^d — supp Γ with $q(A_\alpha)=m$, $m=1, 2, \ldots, r$. The domain Ext Γ is called the exterior of Γ, $\mathrm{Int}\,\Gamma = \bigcup_m \mathrm{Int}_m \Gamma$ is the interior of Γ. The following abbreviations will be used later: $|\Gamma| = |\mathrm{supp}\,\Gamma|$, $V(\Gamma) = |\mathrm{Int}\,\Gamma|$, $V_m(\Gamma) = |\mathrm{Int}_m \Gamma|$.

As an important consequence of the above remarks, to each contour $\Gamma = (M, \varphi(M))$ there corresponds a unique configuration φ_Γ in the following way: $\varphi_\Gamma = \psi_q$ on Ext Γ if $q(\mathrm{Ext}\,\Gamma) = q$, $\varphi_\Gamma(M) = \varphi(M)$, further $\varphi_\Gamma = \psi_m$ on $\mathrm{Int}_m \Gamma$, i.e. φ_Γ is the configuration having Γ as its only contour. If $q(\mathrm{Ext}\,\Gamma) = q$, then we say that Γ, is a contour with boundary condition $\psi_q \in g(\mathscr{H}_0)$; the superscript q in Γ^q indicates that Γ^q is a contour with boundary condition ψ_q. Of course, $\varphi_{\Gamma^q} = \psi_q$ (a.s.). The set of all contours Γ^q with boundary condition $\psi_q \in g(\mathscr{H}_0)$ is denoted by $\mathscr{C}_q$.

If Γ is a contour and $V \subset Z^d$, then $\Gamma \subset V$ means that supp $\Gamma \subset V$, Int $\Gamma \subset V$ and $d(\mathrm{supp}\,\Gamma, V) > 1$. Suppose that Γ and Γ' are contours with pairwise distant supports, then supp $\gamma' \subset$ Int Γ implies $\Gamma' \subset$ Int Γ and excludes $\Gamma \subset$ Int Γ', while supp $\Gamma' \subset$ Ext Γ and supp $\gamma \subset$ Ext Γ' imply that $\Gamma' \subset$ Ext γ and $\Gamma \subset$ Ext Γ'. Therefore, the mutual position of Γ, Γ' with $d(\mathrm{supp}\,\Gamma, \mathrm{supp}\,\Gamma') > 1$ is characterized by exactly one of the following three possibilities. (i) $\Gamma' \subset$ Int Γ, (ii) $\Gamma \subset$ Int Γ', (iii) $\Gamma' \subset$ Ext Γ and $\Gamma \subset$ Ext Γ'. Moreover, if Γ'' is a third contour and $\Gamma'' \subset$ Int Γ', $\Gamma' \subset$ Int Γ, then $\Gamma'' \subset$ Int Γ; i.e. the relation $\Gamma' \subset$ Int Γ defines a partial order in any set of contours with pairwise distant supports. Let us remark that if Γ^q and Γ are contours of the same configuration, then $\Gamma^q \subset \mathrm{Int}_m \Gamma$ is not possible for $m \neq q$.

The description of pure phases at large values of β will be based on the statistical properties of external contours.

Definition 2.6. *A contour Γ of a configuration φ is called an external contour of φ if $\Gamma \subset$ Ext Γ' holds for any other contour Γ' of φ; i.e. the external contours of φ are the maximal elements in the set of contours of φ with respect to the above defined partial order.*

If $\partial\varphi$ is finite, then the set of external contours of φ is a well-defined, nonempty set. In this case all external contours have the same boundary condition and any non-external contour is contained in the interior of one of the external contours. The situation, however, is more complicated if $|\partial\varphi| = \infty$.

The proofs of the forthcoming theorems of this chapter are all based on the properties of the following versions of the statistical sum.

Definition 2.7. *Let $\Omega(\Gamma^q)$ denote the set of configurations $\varphi = \psi_q$ (a.s.) such that $\Gamma^q \in \mathscr{C}_q$ is the only external contour of φ. Then*

$$\Theta(\Gamma^q | \beta\mathscr{H}) = \sum\nolimits_{\varphi \in \Omega(\Gamma^q)} \exp(-\beta\mathscr{H}(\varphi | \psi_q))$$

is called the crystal statistical sum for Γ^q.

Definition 2.8. *Let $V \subset Z^d$ denote a finite set, and $\Omega_q(V)$ the set of configurations $\varphi = \psi_q$ (a.s.) such that $\Gamma \subset V$ whenever Γ is a contour of φ. Then*

$$\Theta_q(V|\beta\mathscr{H}) = \sum_{\varphi \in \Omega_q(V)} \exp\{-\beta\mathscr{H}(\varphi|\psi_q)\}$$

is called the dilute statistical sum in V.

$\Theta(\Gamma^q|\beta\mathscr{H})$ and $\Theta_q(V|\beta\mathscr{H})$ are connected by the following system of recurrence equations.

Lemma 2.3. *For each $V \subset Z^d$ we have*

$$\Theta_q(V|\beta\mathscr{H}) = \sum \prod_{i=1}^{n} \Theta(\Gamma_i^q|\beta\mathscr{H}), \tag{2.17}$$

where the sum is over the set of all families $\{\Gamma_1^q, \ldots, \Gamma_n^q\}$ of external contours in V; i.e. $\Gamma_i^q \subset V$, $i=1, 2, \ldots, n$, and $\Gamma_i^q \subset \text{Ext}\,\Gamma_j^q$, $\Gamma_j^q \subset \text{Ext}\,\Gamma_i^q$ if $i \neq j$. (The value of the empty product is 1.)

Further, for each contour $\Gamma^q \in \mathscr{C}_q$ we have

$$\Theta(\Gamma^v|\beta\mathscr{H}) = \exp[-\beta\mathscr{H}(\Gamma^q)] \prod_{m=1}^{r} \Theta_m(\text{Int}_m\,\Gamma^q|\beta\mathscr{H}), \tag{2.18}$$

where $\mathscr{H}(\Gamma^q) = \mathscr{H}(\varphi_{\Gamma^q}|\psi_q)$ denotes the relative energy of Γ^q.

Lemma 2.4. *Suppose that we are given a real function $\mathscr{Z}(\Gamma^q)$ on $\mathscr{C}_q$ for each $q=1, 2, \ldots, r$, and $\mathscr{Z}_q(V)$ is defined for each finite $V \subset Z^d$ by* (2.17) *with $\mathscr{Z}(\Gamma^q)$ in place of $\Theta(\Gamma^q|\beta\mathscr{H})$. If $\mathscr{Z}(\Gamma^q)$ and $\mathscr{Z}_q(V)$ satisfy* (2.18), *then $\mathscr{Z}(\Gamma^q) = \Theta(\Gamma^q|\beta\mathscr{H})$ on $\mathscr{C}_q$ for each q.*

These trivial lemmas play a central role in the proof of the Main Theorem. Namely, if we are given the functions $\mathscr{H}(\Gamma^q)$, $q=1, 2, \ldots, r$, then (2.17)–(2.18) turns out to be a system of recurrence equations. Because of Peierls's condition, at large values of β this system of equations can be reformulated in terms of a contraction operator in an appropriately chosen Banach space of functions on $\bigcup \mathscr{C}_q$. Proofs will be based on this contraction property of (2.17)–(2.18).

7. Contour models

At large values of β, the solution $\Theta(\Gamma^q|\beta\mathscr{H})$ of the above system of recurrence equations will be investigated by means of contour models. Roughly speaking, a contour model is a family of probability distributions on the set of boundaries consisting of compatible contours with the same boundary condition ψ_q. More exactly, for each class $\mathscr{C}_q$ of contours with boundary condition ψ_q, $q=1, 2, \ldots, r$, we define a contour model. We say that two contours $\Gamma', \Gamma'' \in \mathscr{C}_q$ are compatible if $d(\text{supp}\,\Gamma', \text{supp}\,\Gamma'') > 1$.

Let us remark that compatible contours are not necessarily contours of some configuration, because $\Gamma'' \subset \text{Int}_m\,\Gamma'$ is possible even if $m \neq q$ and $\Gamma', \Gamma'' \in \mathscr{C}_q$. However, external contours of a finite set of compatible contours with the same boundary condition are always external contours of a suitably chosen configuration. In the following definitions and notation we ought to refer to the contour class $\mathscr{C}_q$, $q=1, 2, \ldots, r$, corresponding to the boundary condition ψ_q we have in mind. However, in the abstract theory of contour models this detail is quite unimportant, so we shall keep the notation as economical as possible.

Definition 2.9. *Let $\mathscr{C}$ denote any of the contour classes $\mathscr{C}_q$; $q\in\{1, 2, \ldots, r\}$ is fixed. A finite or countable set ∂ of pairwise compatible contours $\Gamma\in\mathscr{C}$ is called a boundary. The set of all boundaries ∂ is denoted by $\mathscr{D}$; the empty set also belongs to $\mathscr{D}$.*

In what follows, $[\partial]$ denotes the set of contours $\Gamma'\in\mathscr{C}$ for which there exists at least one $\Gamma\in\partial$ such that Γ and Γ' are not compatible.

If $\partial\in\mathscr{D}$ and $V\subset Z^d$, then $\partial\subset V$ means that $\Gamma\subset V$ whenever $\Gamma\in\partial$. The indicator of the set $\{\partial\in\mathscr{D}|\partial\subset V\}$ is denoted by χ_V, i.e. $\chi_V(\partial)=1$ if $\partial\subset V$, and $\chi_V(\partial)=0$ otherwise. Further, $\mathscr{D}_\Gamma=\{\partial\in\mathscr{D}|\Gamma\in\partial\}$ is the set of boundaries containing $\Gamma\in\mathscr{C}$, $\|\partial\|=\sum_{\Gamma\in\partial}|\Gamma|$, and $|\partial|$ denotes the cardinality of $\partial\in\mathscr{D}$, i.e. $|\partial|$ is the number of contours in ∂.

Suppose now that $F=F(\Gamma)$ is a real function defined on $\mathscr{C}$ such that

$$\|F\| = \sup_{\Gamma\in\mathscr{C}}\frac{|F(\Gamma)|}{|\Gamma|} < \infty .$$

Such functions are usually called contour functionals. If there exists a $\tau>0$ such that $F(\Gamma)\geqq\tau|\Gamma|$ for each $\Gamma\in\mathscr{C}$, then we say that F is a τ-functional; the additive extension

$$F(\partial) = \sum_{\Gamma\in\partial} F(\Gamma), \quad \partial\in\mathscr{D}; \quad F(\emptyset) = 0$$

of F to $\mathscr{D}$ satisfies $F(\partial)\geqq\tau\|\partial\|$ for each $\partial\in\mathscr{D}$; this relation is analogous to Peierls's condition for lattice models. A contour functional F is periodic if there exists a subgroup $\hat{Z}\subset Z^d$ of finite index such that $F(T_x\Gamma)=F(\Gamma)$ for each $x\in\hat{Z}$, where

$$T_x\Gamma = (T_x M, T_x\varphi(M))$$

if $\Gamma=(M, \varphi(M))$, and $T_x M=\{y\in Z^d|y=x+z, z\in M\}$.

Definition 2.10. *Let F denote a τ-functional defined on $\mathscr{C}$. Then the crystal and the dilute statistical sums are given by*

$$\mathscr{Z}(\Gamma|F) = e^{-F(\Gamma)} \sum_{\partial\subset\operatorname{Int}\Gamma} e^{-F(\partial)},$$

and

$$\mathscr{Z}(V|F) = \sum_{\partial\subset V} e^{-F(\partial)}$$

respectively, where $V\subset Z^d$ is a finite set, while $\Gamma\in\mathscr{C}$. Then the family P_V of probability distributions on $\mathscr{D}$ given by

$$P_V(\partial) = \chi_V(\partial)\frac{1}{\mathscr{Z}(V|F)} e^{-F(\partial)}, \quad \partial\in\mathscr{D}$$

is called a contour mrdel.

In case of contour models, Lemma 2.3 has the following form.

Lemma 2.5. *The dilute statistical sum can be expressed in terms of the crystal statistical sum as*

$$\mathscr{Z}(V|F) = \sum \prod_{\Gamma\in\partial} \mathscr{Z}(\Gamma|F), \tag{2.19}$$

where the sum is over boundaries $\partial \subset V$ consisting only of their own external contours. For the crystal statistical sum we have

$$\mathscr{Z}(\Gamma|F) = e^{-F(\Gamma)} \prod\nolimits'_{m=1} \mathscr{Z}(\mathrm{Int}_m\, \Gamma|F). \tag{2.20}$$

Contour models form a very peculiar class of interacting systems. Namely, the statistical dependence of contours in a contour model is merely due to the restriction that not each set of contours forms a boundary. Therefore if τ is large, i.e. the density of contours is low, then we can hope that contour models have "good" cluster properties.

Definition 2.11. *The correlation function $\varrho_V(\partial)$, $\partial\in\mathscr{D}$, is defined as the P_V-probability of the event that ∂ is a subset of the random boundary, i.e.*

$$\varrho_V(\partial) = \sum_{\tilde{\partial}\supset\partial} P_V(\tilde{\partial}).$$

Lemma 2.6. *The correlation function of a contour model satisfies the following analogue of Peierls's inequality*

$$\varrho_V(\partial) \leqq e^{-F(\partial)},$$

and the Mayer–Montroll-type equations

$$\varrho_V(\partial) = \chi_V(\partial)\, e^{-F(\partial)} \left[1 + \sum_{\partial'\subset[\partial]} (-1)^{|\partial'|} \varrho_V(\partial')\right]. \tag{2.21}$$

Proof. In view of the definition of ϱ_V we have

$$\varrho_V(\partial) = e^{-F(\partial)} \sum_{\partial\cup\partial'\in\mathscr{D}} P_V(\partial') = e^{-F(\partial)} \left[1 - \sum_{\partial\cup\partial'\notin\mathscr{D}} P_V(\partial')\right],$$

whence Peierls's inequality follows directly.

Since $\partial\cup\partial'\notin\mathscr{D}$ if and only if $[\partial]\cap\partial'\neq\emptyset$, i.e. $\partial'\in\bigcup_{\Gamma\in[\partial]}\mathscr{D}_\Gamma$, and furthermore

$$\varrho_V(\partial') = P_V\left(\bigcap_{\Gamma\in\partial'}\mathscr{D}_\Gamma\right),$$

we have

$$\sum_{\partial':\partial\cup\partial'\notin\mathscr{D}} P_V(\partial') = P_V\left(\bigcup_{\Gamma\in[\partial]}\mathscr{D}_\Gamma\right);$$

whence (2.21) follows immediately by the inclusion-exclusion formula for the probability of sums of events. Q.e.d.

The following elementary lemma will be applied several times in the forthcoming proofs. It is especially useful in estimating sums over $[\partial]$.

Lemma 2.7. *Let* $a = \max\{d \ln(2s+1), \ln|\Phi|+3^d\}$, *where* $s>1$ *is the parameter specifying the class* $\mathbf{H}_s(\hat{Z})$ *of Hamiltonians we consider. Then*

$$\left|\{\Gamma \in \mathscr{C} \mid d(x, \operatorname{supp}\Gamma \cup \operatorname{Int}\Gamma) \leqq 1,\ |\Gamma| = n\}\right| \leqq e^{an}$$

holds for each $x \in Z^d$.

Proof. First we estimate the number

$$C_n = \left|\{\Gamma \in \mathscr{C} \mid d(x, \operatorname{supp}\Gamma) \leqq 1,\ |\Gamma| = n\}\right|.$$

Since the cardinality of the set of finite configurations $\varphi(\operatorname{supp}\Gamma)$ is certainly less than $|\Phi|^n$, it is enough to count the set of connected sets M such that $|M|=n$ and $d(x, M) \leqq 1$. For this set

$$k_i(M) = \{y \in M \mid \|y-x\| = i\} \quad \text{and} \quad k_i = |k_i(M)|$$

for $i=0, 1, 2, \ldots$; since M is connected,

$$M = \bigcup_{i=0}^{n+1} k_i(M) \quad \text{and} \quad n = \sum_{i=0}^{n+1} k_i,$$

further $k_j=0$, whenever $j>i>0$ with $k_i=0$. On the other hand, $k_{i+1}(M)$ is necessarily a subset of $\partial\overline{k_i(M)} = \{y \mid d(y, k_i(M)=1\}$, and $|\partial k_i(M)| \leqq k_i 3^d$. Thus $k_{i+1}(M)$ can be chosen at most $2^{k_i 3^d}$ different ways if $k_i(M)$ is fixed, while $k_0(M)$ is uniquely determined by the value of k_0. Therefore, if the numbers k_i are fixed, then M can be chosen in at most

$$\prod_{i=0}^{n+1} 2^{k_i 3^d} = \exp(n3^d \ln 2)$$

different ways. However, the number of decompositions $n = \sum_{i=0}^{n+1} k_i$ described above is less than 2^{n+1}, so

$$C_n \leqq |\Phi|^n 2^{n+1} \exp(n3^d \ln 2).$$

To prove the final inequality, consider a half-line L^+ parallel to one of the coordinate axes of Z^d and starting from x. Let $x(\Gamma)$ denote the first point z on L^+ such that $d(z, \operatorname{supp}\Gamma) \leqq 1$. Since $\operatorname{supp}\Gamma$ separates $\operatorname{Int}\Gamma$ from $\operatorname{Ext}\Gamma$, $z(\Gamma)$ is well defined. In particular, $z(\Gamma)=x$ if $x \notin \operatorname{Int}\Gamma$, while $z(\Gamma)$ can be chosen in at most n different ways if $x \in \operatorname{Int}\Gamma$. Therefore $n\,C_n$ is a bound for the number of contours we consider. Since $n \geqq 9$,

$$n\,|\Phi|^n 2^{n+1} \exp(n3^d \ln 2) \leqq e^{an}$$

if $a \geqq \ln|\Phi|+3^d$. Q.e.d.

The study of correlation functions ϱ_V will be based on the system (2.21) of correlation equations. The methods we apply here are closely related to the methods developed for the study of Gibbs distributions at low densities and at small values of β. The basic problem is that of the behaviour of ϱ_V as $V \to \infty$. It will be solved in the next section under the assumption that the contour model is given by a τ-functional F and that τ is large enough.

8. Correlation functions of infinite contour models

The right-hand side of (2.21) will be considered as a linear operator on an appropriately chosen Banach space of boundary functionals. A real function $\xi=\xi(\partial)$ defined on the set $\mathscr{D}_0=\{\partial\in\mathscr{D}|\|\partial\|<\infty\}$ of finite boundaries is called a boundary functional if

$$\|\xi\| = \sup_{\partial\in\mathscr{D}_0}\frac{|\xi(\partial)|}{\|\partial\|}<\infty.$$

The set $\mathscr{B}$ of boundary functionals obviously forms a Banach space; now we introduce a family of Banach spaces. For each finite subset W of Z^d and for each τ-functional F with $\tau>3a$ we define the norm $\|\xi\|_W$ of boundary functionals ξ by

$$\|\xi\|_W = \sup_{\partial\in\mathscr{D}_0}\frac{|\xi(\partial)|}{\exp\left[a\|\partial\|-F(\partial)+(a-\tau)d(\partial,\overline{W})\right]}, \tag{2.22}$$

where $\overline{W}=Z^d-W$. The constant is the same as in Lemma 2.7, while

$$d(\partial,\overline{W}) = \min_{\Gamma\in\partial} d(\operatorname{supp}\Gamma,\ \overline{W}).$$

It is easy to check that any of these norms defines a Banach space $\mathscr{B}_W=\{\xi\in\mathscr{B}|\|\xi\|_W<$ $<\infty\}$ of boundary functionals. The right-hand side of (2.21) defines an operator A acting on spaces of boundary functionals ξ as

$$(A\xi)(\partial) = e^{-F(\partial)}\sum_{\partial'\subset[\partial]}(-1)^{|\partial'|}\xi(\partial'). \tag{2.23}$$

The boundary functionals $\chi_V=\chi_V(\partial)$ and $e^{-F}=e^{-F(\partial)}$ can also be considered as multiplicative operators in $\mathscr{B}$. Using the above abbreviations, (2.21) turns into

$$\xi = \chi_V e^{-F}+\chi_V A\chi_V\xi, \tag{2.24}$$

where V is an arbitrary finite subset of Z^d. The correlation functions are also boundary functionals and ϱ_V satisfies (2.24) if V is the same as in ϱ_V. The solution of the equation

$$\xi = e^{-F}+A\xi \tag{2.25}$$

will be interpreted as the correlation function of the contour model in infinite volume. More exactly, we prove that if τ is large enough, then the limit $\varrho=\lim_{V\to\infty}\varrho_V$ exists and ϱ is the unique solution of (2.25). The basic tool of the proof is

Lemma 2.8. *If F is a τ-functional with $\tau\geqq 4a$, then $\|A\|_W\leqq e^{-a}$ for each finite $W\subset Z^d$.*

Proof. Assume that $\|\xi\|_W\leqq 1$. Then $\tau>a$ and $F(\partial')\geqq\tau\|\partial'\|$ imply that

$$|(A\xi)(\partial)| \leqq e^{-F(\partial)}\sum_{\partial'\subset[\partial]}\exp\left[(a-\tau)\left(\|\partial'\|+d(\partial',\overline{W})\right)\right]. \tag{2.26}$$

First we show that $d(\partial', \overline{W}) \geqq d(\partial, \overline{W}) - \frac{1}{2}\|\partial'\|$ whenever $\partial' \subset [\partial]$. Indeed, there exist contours $\Gamma' \in \partial'$, $\Gamma \in \partial$ and points $x' \in \operatorname{supp} \Gamma'$, $z \in \overline{W}$, $x \in \operatorname{supp} \Gamma$ such that $d(\partial', W) = d(\operatorname{supp} \Gamma', \overline{W}) = |x' - z|$ and $d(\operatorname{supp} \Gamma, \operatorname{supp} \Gamma') = |x - y'| \leqq 1$. Consequently

$$d(\partial, \overline{W}) \leqq |x-z| \leqq |x-y'| + |y'-x'| + |x'-z| \leqq$$
$$\leqq 1 + \operatorname{diam} \operatorname{supp} \Gamma' + d(\partial', \overline{W}).$$

On the other hand, $\operatorname{supp} \Gamma'$ is a connected set and it is a union of cubes $W_s(x)$ with $s \geqq 2$, so $1 + \operatorname{diam} \operatorname{supp} \Gamma' \leqq \frac{1}{2}|\Gamma'| \leqq \frac{1}{2}\|\partial'\|$, and (2.26) turns into

(2.27) $$|(A\xi)(\partial)| \leqq \exp[(a-\tau)\, d(\partial, W) - F(\partial)] \sum_{\partial' \subset [\partial]} \exp\left[\frac{1}{2}(a-\tau)\|\partial'\|\right].$$

Now we estimate the sum in (2.27); the following counting procedure will be used. First we select $k = |\partial'|$ different points $x_1, x_2, \ldots, x_k$ from the set

$$M = \bigcup_{\Gamma \in \partial} \operatorname{supp} \Gamma,$$

and then we choose k different contours $\Gamma'_1, \Gamma'_2, \ldots, \Gamma'_k$ such that $d(x_i, \operatorname{supp} \Gamma'_i) \leqq 1$ for each i. Since any contour is a union of cubes $W_s(x)$ with $s \geqq 2$, and furthermore the contours $\Gamma' \in \partial'$ are distant, each $\partial' \subset [\partial]$ can be obtained as $\partial' = \{\Gamma'_1, \Gamma'_2, \ldots, \Gamma'_k\}$ by the above procedure, and $|\partial'| \leqq \|\partial\|$ if $\partial' \subset [\partial]$. The number of contours Γ'_i such that $d(x_i, \Gamma'_i) \leqq 1$ and $|\Gamma'_i|$ is fixed, has been estimated in Lemma 2.7. Thus

(2.28) $$\sum_{\partial' \subset [\partial]} e^{-c\|\partial'\|} \leqq \sum_{k=1}^{\|\partial\|} C^k_{\|\partial\|} \prod_{i=1}^k \Big(\sum_{d(x_i, \Gamma_i) \leqq 1} e^{-c|\Gamma_i|}\Big) \leqq$$
$$\leqq \sum_{k=0}^{\|\partial\|} C^k_{\|\partial\|} \Big[\sum_{n=1}^{\infty} e^{(a-c)n}\Big]^k = [1 - e^{a-c}]^{-\|\partial\|}$$

for each $c \geqq a$; C^k_n denotes the binomial coefficient. Comparing (2.27) and (2.28), we obtain

(2.29) $$|(A\xi)(\partial)| \leqq \exp[a\|\partial\| - F(\partial) + (a-\tau)\, d(\partial, \overline{W})] \left[\frac{e^{-a}}{1 - \exp\left[\frac{3}{2}a - \frac{1}{2}\tau\right]}\right]^{\|\partial\|}.$$

Since $a \geqq 9$ and $\|\partial\| \geqq 9$, (2.29) implies the statement directly. Q.e.d.

Now we are in a position to describe the weak limit

(2.30) $$P(\cdot|F) = \lim_{V \to \infty} P_V(\cdot|F).$$

$P_V(\cdot|F)$ is a probability measure on the σ-algebra of subsets of $\mathscr{D}$ generated by the cylinder sets $\mathscr{D}_\Gamma$, $\Gamma \in \mathscr{C}$. As usually, $V \to \infty$ means that $V = V_n$ is a sequence of finite subsets V_n of Z^d, and $\lim d(x, \overline{V}_n) = 0$ for each $x \in Z^d$.

Proposition 2.1. *Let F denote a τ-functional with $\tau \geqq 4a$, and a the absolute constant introduced in Lemma 2.7. Then for each finite boundary $\partial \in \mathscr{D}_0$ the infinite volume*

correlation function $\varrho(\partial)=\lim_{V\to\infty}\varrho_V(\partial)$ *exists;* ϱ *satisfies Peierls's inequality*

$$\varrho(\partial)\leqq e^{-F(\partial)}$$

and

$$|\varrho(\partial)-\varrho_V(\partial)|\leqq \exp\left[(a-\tau)(\|\partial\|+d(\partial,\bar V))\right]$$

if $\partial\subset V$.

Proof. $\varrho(\partial)$, $\partial\in\mathscr{D}_0$ is defined as the solution of the equation $\varrho=e^{-F}+A\varrho$; we have to prove that $\lim_{V\to\infty}\varrho_V=\varrho$. To prove these statements, first we consider A as an operator on the Banach space $\mathscr{B}_\varnothing$ of boundary functionals, $\mathscr{B}_\varnothing$ corresponds to the empty set $W=\varnothing$. Since $\|A\|_\varnothing<1$ and $\|e^{-F}\|\leqq e^{-a}<+\infty$, ϱ is uniquely defined and

$$\varrho=\sum_{n=0}^{\infty}A^n e^{-F}. \tag{2.31}$$

On the other hand, $\varrho_V=\chi_V e^{-F}+\chi_V A\chi_V\varrho_V$ and χ_V is a projection into $\{\partial\in\mathscr{D}_0|\partial\subset V\}$, so

$$\chi_V\varrho-\varrho_V=\chi_V A(\varrho-\chi_V\varrho)+\chi_V A(\chi_V\varrho-\varrho_V);$$

i.e. $\chi_V\varrho-\varrho_V$ is again a fixed point and

$$\chi_V\varrho-\varrho_V=\sum_{n=0}^{\infty}(\chi_V A)^n\eta_V,$$

where $\eta_V=\chi_V A\varrho-\chi_V A\chi_V\varrho$. The above series converges in any of the Banach spaces $\mathscr{B}_w$ in which the norm of η_V is finite; we estimate $\|\eta_V\|_V$ as follows. Observe that $(\varrho-\chi_V\varrho)(\partial')=0$ if $\partial'\subset V$ and $(\varrho-\chi_V\varrho)(\partial')=\varrho(\partial')$ if $\partial'\not\subset V$. In the second case $\partial'\subset[\partial]$ implies that $|\Gamma'|\geqq d(\partial,\bar V)+1$ holds for at least one contour $\Gamma'\in\partial'$, therefore repeating the argument that resulted in (2.29) we obtain

$$|\eta_V(\partial)|\leqq e^{-F(\partial)}\sum_{\partial'\subset[\partial]}|\varrho-\chi_V\varrho(\partial')|\leqq \tag{2.32}$$

$$\leqq\|\varrho\|_\varnothing e^{-F(\partial)+(a-\tau)d(\partial,\bar V)}\sum_{k=0}^{\|\partial\|}C^k_{\|\partial\|}\left[\sum_{n=1}^{\infty}e^{(2a-\tau)n}\right]^k=$$

$$=\|\varrho\|_\varnothing\exp\left[a\|\partial\|-F(\partial)+(a-\tau)\,d(\partial,\bar V)\right]\left[\frac{e^{-a}}{1-e^{2a-\tau}}\right]^{\|\partial\|}$$

i.e. $\|\eta_V\|_V\leqq e^{-a}\|\varrho\|_\varnothing\leqq e^{-a}\|e^{-F}\|_\varnothing(1-\|A\|_\varnothing)^{-1}$ if $\tau>4a$, whence it follows that

$$\|\chi_V\varrho-\varrho_V\|_V\leqq\|\eta_V\|_V(1-\|A\|_V)^{-1}<1;$$

which implies all the statements of the proposition directly. Q.e.d.

Corollary. *For each* $x\in Z^d$ *we have*

$$P\big(x\in\textstyle\bigcup_{\Gamma\in\partial}(\operatorname{supp}\Gamma\cup\operatorname{Int}\Gamma)|F\big)\leqq e^{-\tau}$$

provided that $\tau\geqq 4a$.

Proof. In view of Peierls's inequality and Lemma 2.7, the probability on the left-hand side is bounded by

$$\sum_{\Gamma: x\in \operatorname{supp}\Gamma \cup \operatorname{Int}\Gamma} e^{-\tau|\Gamma|} \leqq \sum_{n=9}^{\infty} e^{(a-\tau)n} \leqq e^{-\tau}.$$

Q.e.d.

This corollary formulates the intuitive fact that the density of contours is low if τ is large. To obtain a deeper insight into the structure of infinite contour configurations, we also investigate the distribution of external contours.

Consider an infinite contour model and let $\Theta(\partial)$ denote the set of external contours of the random boundary $\partial\in\mathscr{D}$, i.e.

$$\Theta(\partial) = \{\tilde{\Gamma}\in\partial | \tilde{\Gamma} \subset \operatorname{Ext}\Gamma \text{ if } \Gamma\in\partial, \Gamma\neq\tilde{\Gamma}\}.$$

Remember that $\Theta(\partial)=\emptyset$ is possible if ∂ is an infinite configuration. Elements of the set $\tilde{\mathscr{D}}=\{\partial\in\mathscr{D}|\partial=\Theta(\partial)\}$ are called external boundaries, $\tilde{\mathscr{D}}_0=\tilde{\mathscr{D}}\cap\mathscr{D}_0$. The correlation functions $\tilde{\varrho}_V=\tilde{\varrho}_V(\Theta)$ and $\varrho=\varrho(\Theta)$ of external boundaries are defined as

$$\tilde{\varrho}_V(\Theta) = P_V(\Theta\subset\Theta(\partial)), \quad \Theta\in\tilde{\mathscr{D}}_0$$

and

$$\tilde{\varrho}(\Theta) = P(\Theta\subset\Theta(\partial)|F), \quad \Theta\in\tilde{\mathscr{D}}_0.$$

We shall see that they also satisfy some correlation equations similar to (2.25).

Proposition 2.2. *Let F denote a τ-functional with $\tau\geqq 4a$,* and *$P(\cdot|F)$ the corresponding infinite contour model described in Proposition* 2.1. *Then for almost every $\partial\in\mathscr{D}$ the set $\Theta(\partial)$ of external contours of ∂ is complete in the sense that $\Theta(\partial)\neq\emptyset$, and for each $\Gamma\in\partial$ there exists exactly one $\tilde{\Gamma}\in\Theta(\partial)$ such that $\tilde{\Gamma}\subset\operatorname{Int}\Gamma$. Further,* $\lim_{V\to\infty}\tilde{\varrho}_V(\Theta)=\tilde{\varrho}(\Theta)$ *for each $\Theta\in\tilde{\mathscr{D}}_0$.*

$$\varrho(\Theta) \leqq e^{-\tau\|\Theta\|}$$

and $\tilde{\varrho}$ satisfies the following exponential mixing condition. Suppose that $\Theta', \Theta''\in\tilde{\mathscr{D}}_0$ are finite external boundaries such that even $\Theta'\cup\Theta''\in\tilde{\mathscr{D}}_0$, then

$$|\tilde{\varrho}(\Theta'\cup\Theta'')-\tilde{\varrho}(\Theta')\tilde{\varrho}(\Theta'')| \leqq \exp\left[(a-\tau)(\|\Theta'\|+\|\Theta''\|+d(\Theta',\Theta''))\right],$$

where $d(\Theta',\Theta'')=\min_{\Gamma''\in\Theta''} d(\Theta', \operatorname{supp}\Gamma'')$.

Proof. The first statement is a direct consequence of Lemma 2.7 and of Peierls's inequality (see Proposition 2.1). Indeed, let $x\in Z^d$ be fixed and suppose that there exists an infinite sequence $\Gamma_n\in\partial$ of contours of the arbitrary boundary $\partial\in\mathscr{D}$ such that $x\in\operatorname{Int}\Gamma_n$ and $\Gamma_n\subset\operatorname{Int}\Gamma_{n+1}$ for each n. Then

$$\sum_{n=1}^{\infty} P(x\in\operatorname{Int}\Gamma_n) \leqq \sum_{n=1}^{\infty} e^{(a-\tau)n} < +\infty$$

as $|\Gamma_{n+1}|\geqq|\Gamma_n|\geqq n$. Thus in view of the Borel–Cantelli lemma, the above situation is possible only with probability 0, i.e. $\Theta(\partial)$ is complete for almost every $\partial\in\mathscr{D}$.

The study of $\tilde{\varrho}$ follows the lines of the proofs of Lemma 2.8 and Proposition 2.1; we repeat an argument only if a modification is needed.

First we consider correlation functions $\tilde{\varrho}_V(\Theta)$, $\Theta \subset V$, $\Theta \in \tilde{\mathscr{D}}_0$ in finite volumes. By definition,

$$\tilde{\varrho}_V(\Theta) = P_V(\Theta \subset \Theta(\partial)) = e^{-F(\Theta)} \sum_{\partial' : \Theta \subset \Theta(\partial' \cup \Theta)} P_V(\partial') =$$
$$= e^{-F(\Theta)} [1 - P_V(\bigcup_{\Gamma \in [\tilde{\Theta}]} \tilde{\mathscr{D}}_\Gamma)],$$

where $\tilde{\mathscr{D}}_\Gamma = \{\partial \in \mathscr{D} | \Gamma \in \Theta(\partial)\}$, and

$$[\tilde{\Theta}] = [\Theta] \cup \{\Gamma \in \mathscr{C} | \tilde{\Gamma} \subset \operatorname{Int} \Gamma \text{ for some } \tilde{\Gamma} \in \Theta\}.$$

Then $\tilde{\varrho}_V$ satisfies the modified correlation equation

$$\xi = \chi_V e^{-F} + \chi_V \tilde{A} \chi_V \xi$$

where the operator $\tilde{A}$ acts on the space of functions on $\tilde{\mathscr{D}}_0$ as

$$(\tilde{A}\xi)(\Theta) = e^{-F(\Theta)} \sum_{\Theta' \subset [\tilde{\Theta}]} (-1)^{|\Theta'|} \xi(\Theta'); \tag{2.33}$$

the equation for $\tilde{\varrho}$ reads $\xi = e^{-F} + \tilde{A}\xi$.

The next step is to show that $\|\tilde{A}\|_W < e^{-a}$ for each finite $W \subset Z^d$. For this only the counting procedure in the proof of Lemma 2.8 should be modified, as follows. Select first $k = |\Theta'|$ different points $x_1, x_2, \ldots, x_k$ from the set

$$\tilde{M} = \bigcup_{\Gamma \in \Theta} (\operatorname{supp} \Gamma \cup \operatorname{Int} \Gamma),$$

and select k different contours $\Gamma_1, \ldots, \Gamma_k$ such that $d(x_i, \operatorname{supp} \Gamma \cup \operatorname{Int} \Gamma) \leqq 1$. Since each external boundary $\Theta' \subset [\tilde{\Theta}]$ can be obtained in this way, $\|\tilde{A}\|_w < e^{-a}$ follows from Lemma 2.7 in the same way as $\|A\|_w < e^{-a}$ has been obtained. Having this combinatorial result it follows, in the same way as in the case of Proposition 2.1, that $\tilde{\varrho}$ is the unique solution of $\xi = e^{-F} + \tilde{A}\xi$. Furthermore

$$|\tilde{\varrho}_V(\Theta) - \tilde{\varrho}(\Theta)| \leqq \exp[(a - \tau)(\|\Theta\| + d(\Theta, \bar{V}))], \tag{2.34}$$

whenever $\Theta \in \tilde{\mathscr{D}}_0$ and $\Theta \subset V$.

Now we are in a position to prove the mixing property of $\tilde{\varrho}$. Let

$$M'' = \bigcup_{\Gamma \in \Theta''} (\operatorname{supp} \Gamma \cup \operatorname{Int} \Gamma)$$

and observe that for $\Theta' \cup \Theta'' \subset V$ we have the identity

$$\frac{\tilde{\varrho}_V(\Theta' \cup \Theta'')}{\tilde{\varrho}_V(\partial'')} = \tilde{\varrho}_V(\Theta' | \Theta'') = \tilde{\varrho}_{V - M''}(\Theta'),$$

and, in view of (2.34),

$$|\tilde{\varrho}_{V - M''}(\Theta') - \tilde{\varrho}(\Theta')| \leqq \exp[(a - \tau)(\|\Theta'\| + d(\Theta', \bar{V} \cup M''))].$$

Since $\tilde{\varrho}_V(\Theta'') \leqq \varrho_V(\Theta'') \leqq e^{-F\|\Theta''\|} e^{-\tau\|\Theta''\|}$, it follows that

$$|\tilde{\varrho}_V(\Theta' \cup \Theta'') - \tilde{\varrho}_V(\Theta')\tilde{\varrho}_V(\Theta'')| \leqq \exp\left[(a-\tau)\left(\|\Theta'\| + \|\Theta''\| + d(\Theta', V-M'')\right)\right],$$

so that letting $V \to \infty$ we obtain the statement. Q.e.d.

As we shall see soon, some of the limit Gibbs distributions $P^{(q}_{\beta,\mu}$ are associated with a τ-functional F_q defined on $\mathscr{C}_q$ in such a way that $\Theta(\Gamma^q|\beta\mathscr{H}_\mu) = \mathscr{Z}(\Gamma^q|F^q)$ for each $\Gamma^q \in \mathscr{C}^q$. This means that the joint distribution of external contours in the lattice system is the same as in the corresponding infinite contour model; i.e. such a $P^{(q)}_{\beta,\mu}$ is really a pure phase.

9. Surface tension in contour models

In this section we investigate the boundary term of the logarithm of the statistical sum in contour models. If F is a periodic τ-functional and V goes to infinity in the Van Hove sense, i.e. $\dfrac{|\partial V|}{|V|} \to 0$, then the limit

$$s(F) = \lim_{V\to\infty} \frac{1}{|V|} \ln \mathscr{Z}(V|F)$$

exists and is usually referred to as the pressure. The boundary term Δ of $\ln \mathscr{Z}(V|F)$ is defined as $\Delta(V|F) = \ln \mathscr{Z}(V|F) - s(F)|V|$; we shall show that it is really of order $|\partial V|$. Applying (2.20) we obtain a similar decomposition

$$\ln \mathscr{Z}(\Gamma|F) = s(F)|\mathrm{Int}\,\Gamma| - F(\Gamma) + \nabla(\Gamma|F), \tag{2.35}$$

for the crystal statistical sum, where

$$\nabla(\Gamma|F) = \sum_{m=1}^{r} \Delta(\mathrm{Int}_m\,\Gamma|F).$$

The quantity

$$\chi(\Gamma|F) = \frac{1}{|\Gamma|}[F(\Gamma) - \nabla(\Gamma|F)]$$

can be interpreted as the surface tension. It is of fundamental importance in the study of phase transitions. For the physical interpretation, see Section 1 and the proofs in Sections 9 and 10 of this chapter.

Proposition 2.3. *Suppose that F is a $\hat{Z}$-periodic τ-functional with $\tau \geqq 4a$, where a is the constant of Lemma 2.7 while $\hat{Z}$ denotes the symmetry group of our space $\mathbf{H}_s(\hat{Z})$ of Hamiltonians; i.e. $\hat{Z}$ is subgroup of Z^d of finite index and $|Z^d/\hat{Z}| < s$. Then the pressure $s(F)$ exists with $0 \leqq s(F) \leqq e^{-\tau}$, and furthermore $|\Delta(V|F)| \leqq e^{-\tau}|\partial V|$.*

Proof. Let $F_t = tF$, $t \geqq 1$; $\varrho_V(\Gamma, t)$ and $\varrho(\Gamma, t)$ denote respectively the correlation functions for $\partial = \{\Gamma\}$ in the corresponding finite and infinite contour models. Then

$$\frac{d}{dt} \ln \mathscr{Z}(V|F_t) = -\sum_{\Gamma \subset V} F(\Gamma)\varrho_V(\Gamma, t)$$

and $\ln \mathscr{Z}(V|F_\infty) = 0$ as the empty boundary also belongs to $\mathscr{D}$. Therefore

$$\ln \mathscr{Z}(V|F) = \int_1^\infty \sum_{\Gamma \subset V} F(\Gamma)\varrho_V(\Gamma, t)\, dt. \tag{2.36}$$

Looking at (2.36), it is not very hard to recognize that

$$s(F) = \frac{1}{N}\int_1^\infty \sum_{\Gamma \in \hat{\mathscr{C}}} F(\Gamma)\varrho(\Gamma, t)\, dt, \tag{2.37}$$

provided that the limit defining the pressure $s(F)$ exists. Here $N \leqq s$ is the cardinality of the factor group $Z^d/\hat{Z}$, and $\hat{\mathscr{C}} \subset \mathscr{C}$ is chosen in such a way that it contains exactly one contour from each class of $\hat{Z}$-congruent contours. (Γ and Γ' are called $\hat{Z}$-congruent if there exists a $z \in \hat{Z}$ such that $\Gamma = T_z\Gamma$.) The boundary term $\Delta(V|F)$ will be estimated by $\Delta_1(V|F) + \Delta_2(V|F)$, where Δ_1 and Δ_2 are defined in (2.38) and (2.39) respectively.

In the following calculation the estimates of Lemma 2.7 and Proposition 2.1 are used.

$$\Delta_1(V|F) = \int_1^\infty \sum_{\Gamma \subset V} F(\Gamma)[\varrho_V(\Gamma, t) - \varrho(\Gamma, t)]\, dt \leqq \tag{2.38}$$

$$\leqq \int_1^\infty \sum_{\Gamma \subset V} F(\Gamma) \exp[a|\Gamma| + tF(\Gamma) + (a - t\tau)\, d(\Gamma, \bar{V})]\, dt =$$

$$= \sum_{\Gamma \subset V} \frac{F(\Gamma)}{F(\Gamma) + \tau d(\Gamma, \bar{V})} \exp[a|\Gamma| - F(\Gamma) + (a-\tau)\, d(\Gamma, \bar{V})] \leqq$$

$$\leqq \sum_{\Gamma \subset V} \exp[(a-\tau)(|\Gamma| + d(\Gamma, V))] \leqq$$

$$\leqq |\partial V| \sum_{k=0}^\infty \sum_{n=9}^\infty (2k+1)^d e^{an} \exp[(a-\tau)(n+k)] \leqq$$

$$\leqq |\partial V| \frac{e^{18a - 9\tau}}{(1 - e^{2a-\tau})^2} \leqq \frac{1}{2} e^{-\tau} |\partial V|.$$

Further

$$\Delta_2(V|F) = \int_1^\infty \sum_{d(\Gamma, \partial V) \leqq N} F(\Gamma)\varrho(\Gamma, t)\, dt \leqq \tag{2.39}$$

$$\leqq \int_1^\infty \sum_{d(\Gamma, \partial V) \leqq N} F(\Gamma) e^{-tF(\Gamma)}\, dt \leqq \sum_{d(\Gamma, \partial V) \leqq N} e^{-\tau|\Gamma|} \leqq$$

$$\leqq |\partial V|(2N+1)^d \sum_{n=9}^\infty e^{(a-\tau)n} \leqq |\partial V| \frac{(2N+1)^d e^{9a-9\tau}}{1 - e^{a-\tau}} \leqq$$

$$\leqq \frac{1}{2} e^{-\tau} |\partial V|.$$

Observe that the sum on the right-hand side of (2.36) can be decomposed in the following way.

$$\ln \mathscr{Z}(V|F) = \int_1^\infty \sum' F(\Gamma)\varrho(\Gamma, t)\,dt + \int_1^\infty \sum'' F(\Gamma)\varrho(\Gamma, t)\,dt +$$

$$+\int_1^\infty \sum_{\Gamma\subset V} F(\Gamma)[\varrho_V(\Gamma, t) - \varrho(\Gamma, t)]\,dt,$$

where the first sum $\sum'$ has been chosen in such a way that exactly $|V|/N$ contours appear in it from each class of $\hat{Z}$-congruent contours, while only those contours Γ appear in $\sum''$ for which $d(\Gamma, \partial V) \leqq N$. Therefore the first term on the right-hand side is just $s(F)|V|$. Thus

$$\left|\ln \mathscr{Z}(V|F) - s(F)|V|\right| \leqq \Delta_2(V|F) + \Delta_1(V|F) \leqq e^{-\tau}|\partial V|.$$

Finally, in a similar way as in (2.39), we obtain from (2.37) that

$$0 \leqq s(F) \leqq \frac{1}{N}\int_1^\infty \sum_{\Gamma\in\hat{\mathscr{C}}} F(\Gamma)\varrho(\Gamma, t)\,dt \leqq$$

$$\leqq \frac{1}{N}\sum_{\Gamma\in\hat{\mathscr{C}}} e^{-F(\Gamma)} \leqq \frac{1}{N}\sum_{n=9}^\infty e^{(a-\tau)n} \leqq \frac{e^{9a-9\tau}}{N(1-e^{a-\tau})} \leqq e^{-\tau}.$$

Q.e.d.

To study some further properties of $\mathscr{Z}(V|F)$, we need a new norm $|||\cdot|||$ in the space of τ-functionals, namely,

$$|||F||| = \sup_{\Gamma\in\mathscr{C}} \frac{|F(\Gamma)|}{(|\Gamma| + V(\Gamma))e^{a\delta(\Gamma)}},$$

where $\delta(\Gamma) = \operatorname{diam}\operatorname{supp}\Gamma$, and a is the same constant as in Lemma 2.7.

Proposition 2.4. *If F and F' are τ-functionals with $\tau \geqq 4a$, then*

$$|s(F) - s(F')| \leqq e^{-\tau}|||F - F'|||.$$

Proof. We have to prove that

$$\left|\ln \mathscr{Z}(V|F) - \ln \mathscr{Z}(V|F')\right| \leqq e^{-\tau}|V|\,|||F - F'|||.$$

We shall apply the Lagrange theorem. Indeed, Peierls's inequality implies that

$$\left|\frac{\partial \ln \mathscr{Z}(V|\bar{F})}{\partial \bar{F}(\Gamma)}\right| = \varrho_V(\Gamma) \leqq e^{-\bar{F}(\Gamma)},$$

where $\bar{F} = tF + (1-t)F'$, $0 \leqq t \leqq 1$, is again a τ-functional with $\tau \geqq 4a$. Therefore as

$\delta(\Gamma)\leqq|\Gamma|$ and $|V(\Gamma)|\leqq|\Gamma|^d$, it follows that

$$\left|\ln \mathscr{Z}(V|F)-\ln \mathscr{Z}(V|F')\right| \leqq \sum_{\Gamma\subset V} e^{-\tau|\Gamma|}\,|F(\Gamma)-F'(\Gamma')| \leqq$$
$$\leqq |||F-F'|||\sum_{\Gamma\subset V} e^{-\tau|\Gamma|}(|\Gamma|+N(\Gamma))e^{\delta(\Gamma)} \leqq$$
$$\leqq |V|\,|||F-F'|||\sum_{n=9}^{\infty}(n+n^d)\,e^{an}\,e^{(a-\tau)n} \leqq$$
$$\leqq |V|\,|||F-F'|||\,e^{-\tau}.$$

Q.e.d.

Definition 2.12. *Let $b\geqq 0$, and let F be a τ-functional. Then*

$$\mathscr{Z}(V|F,b) = \sum_{\partial\subset V} e^{-F(\partial)}\prod_{\Gamma\in\Theta(\partial)} e^{bV(\Gamma)}$$

is called the parametric statistical sum.

It is easy to see that $\mathscr{Z}(V|F,0)=\mathscr{Z}(V|F)$ and

$$\mathscr{Z}(V|F) \leqq \mathscr{Z}(V|F,b) \leqq \mathscr{Z}(V|F)\,e^{b|V|}$$

as $b\geqq 0$ and $0\leqq V(\Gamma)\leqq|V|$ if $\Gamma\subset V$. Further, if F is a $\hat{Z}$-periodic τ-functional with $\tau\geqq 4a$, then Proposition 2.3 implies

$$-b|V|+e^{-\tau}|\partial V| \leqq \ln\mathscr{Z}(V|F,b)-\big(s(F)+b\big)|V| \leqq \leqq e^{-\tau}|\partial V|. \tag{2.40}$$

A more complete description of the properties of the parametric contour statistical sum is given in the following proposition. We investigate its boundary term $\Delta(V|F,b)$ defined by

$$\ln\mathscr{Z}(V|F,b) = \big(s(F)+b\big)|V|+\Delta(V|F,b).$$

Proposition 2.5. *Let F', F'' be $\hat{Z}$-periodic τ-functionals, $\tau\geqq 4a$, $b', b''\geqq 0$; then*

$$\left|\Delta(V|F',b')-\Delta(V|F'',b'')\right| \leqq 2|b-b'|\,|V|+$$
$$+\left(\frac{1}{e^a-1}+e^{-\tau}\right)e^{a\delta(V)}|V|\,|||F'-F''|||.$$

Proof. It is enough to investigate the particular cases $b'=b''$ and $F'=F''$. In the second case we have

$$\left|\frac{\partial\ln\mathscr{Z}(V|F,b)}{\partial b}\right| \leqq |V|,$$

whence

$$\left|\Delta(V|F,b')-\Delta(V|F,b'')\right| \leqq |b'-b''|\,|V|$$

follows directly.

In the first case

$$\left|\Delta(V|F', b) - \Delta(V|F'', b)\right| \leqq |s(F') - s(F'')||V| + \\ + \left|\ln \mathscr{Z}(V|F', b) - \ln \mathscr{Z}(V|F'', b)\right|,$$

where

$$\left|s(F') - s(F'')\right| \leqq e^{-\tau}|||F' - F''|||,$$

so we have to prove that

$$\left|\ln \mathscr{Z}(V|F', b) - \ln \mathscr{Z}(V|F, b)\right| \leqq |V| \frac{e^{a\delta(V)}}{e^a - 1} |||F' - F''|||.$$

For this purpose we introduce the parametric contour model, i.e. a probability distribution $P_V(\partial|F, b)$ for $\partial \subset V$ as

$$P_V(\partial|F, b) = \mathscr{Z}^{-1}(V|F, b) e^{-F(\partial)} \prod\nolimits_{\Gamma \in \Theta(\partial)} e^{bV(\Gamma)}.$$

It is easy to check that

$$\frac{\partial \ln \mathscr{Z}(V|F, b)}{\partial F(\Gamma)} = -\sum\nolimits_{\partial \in \Gamma} \varrho_V(\partial|F, b),$$

whence

$$\ln \mathscr{Z}(V|F', b) - \ln \mathscr{Z}(V|F'', b) = \\ = -\sum\nolimits_{\Gamma \subset V} \sum\nolimits_{\partial \in \Gamma} \varrho_V(\partial|\bar{F}, b)\big(F'(\Gamma) - F''(\Gamma)\big)$$

follows for $\bar{F} = tF' + (1-t)F''$ with some t between 0 and 1. Further,

$$-\sum\nolimits_{\Gamma \subset V} \sum\nolimits_{\partial \in \Gamma} \varrho_V(\partial|\bar{F}, b)\big(F'(\Gamma) - F''(\Gamma)\big) = \\ = -\sum\nolimits_{\partial \subset V} \varrho_V(\partial|\bar{F}, b)\big(F'(\partial) - F''(\partial)\big);$$

consequently

$$\left|\ln \mathscr{Z}(V|F', b) - \ln \mathscr{Z}(V|F'', b)\right| \leqq \\ \leqq |||F' - F''||| \sum\nolimits_{\partial \subset V} \varrho_V(\partial|\bar{F}, b) \sum\nolimits_{\Gamma \in \partial} \big(|\Gamma| + V(\Gamma)\big) e^{a\delta(\Gamma)}.$$

Since ϱ_V is a probability distribution, what has remained to be proved is that

$$\alpha(V) = \max\nolimits_{\partial \subset V} \frac{1}{|V|} \sum\nolimits_{\Gamma \in \partial} \big(|\Gamma| + V(\Gamma)\big) e^{a\delta(\Gamma)} \leqq \frac{e^{a\delta(V)}}{e^a - 1}.$$

Let $\gamma(n) = \max_{\delta(V) \leqq n} \alpha(V)$. Then

$$\sum\nolimits_{\Gamma \in \partial} \big(|\Gamma| + V(\Gamma)\big) e^{a\delta(\Gamma)} \leqq \sum\nolimits_{\Gamma \in \Theta(\partial)} \big(|\Gamma| + V(\Gamma)\big) e^{a\delta(\Gamma)} + \\ + \sum\nolimits_{\Gamma \in \Theta(\partial)} V(\Gamma) \alpha(\operatorname{Int} \Gamma) \leqq |V| \big[\exp\big(a(\delta(V) - 1)\big) + \gamma(\delta(V) - 1)\big]$$

as $\delta(\Gamma) \leqq \delta(V) - 1$ and $\delta(\operatorname{Int} \Gamma) \leqq \delta(V) - 1$, if $\Gamma \subset V$; while

$$\sum\nolimits_{\Gamma \in \Theta(\partial)} \big(|\Gamma| + V(\Gamma)\big) \leqq |V|$$

if $\partial \subset V$. This means that

$$\alpha(V) \leqq \exp a(\delta(V)-1)+\gamma(\delta(V)-1),$$

i.e. $\gamma(n) \leqq e^{a(n-1)}+\gamma(n-1)$, whence

$$\gamma(n) \leqq e^{a(n-1)}+e^{a(n-2)}+\ldots+e^{a}+1=\frac{e^{an}-1}{e^{a}-1}.$$

Q.e.d.

10. Proof of the Main Theorem

The basic idea of the proof is to describe pure phases by means of contour models in such a way that the joint distributions of external contours in the pure phase and in the corresponding contour model coincide. If β is large and $|\mu|$ is small, then for each periodic ground state ψ_q, $q=1, 2, \ldots, r$, there exists a parameter $b_q \geqq 0$ and a τ-functional F_q defined on the set $\mathscr{C}_q$ of contours with boundary condition ψ_q such that $\Theta_q(V|\beta H_\mu)=\mathscr{Z}(V|F_q, b_q)$ holds for each finite $V\subset Z^d$. If b_q happens to be zero, then $\Theta_q(V|\beta\mathscr{H}_\mu)=\mathscr{Z}(V|F_q)$; i.e. the joint distributions of external contours in the pure phase are really the same as in the associated contour model. Therefore the results of Section 8 can be reformulated for the pure phase. Namely, if $b_q=0$, then the corresponding limit Gibbs distribution $P^{(q)}_{\beta,\mu}$ exists, and the density of contours in this pure phase is low if β is large; so that $P^{(q)}_{\beta,q}$ describes small local distortions of the ground state ψ_q. Moreover, the mixing property stated in Proposition 2.2. implies that $P^{(q)}_{\beta,\mu}$ is indecomposable if $b_q=0$.

Now we turn to the details of the proof of the main result of this chapter. As earlier, contours are defined with respect to s and the set $g(\mathscr{H}_0)$ of periodic ground states of the basic Hamiltonian $\mathscr{H}_0$; the positive integer s is just the parameter of the space $\boldsymbol{H}_s(\hat{Z})$ of Hamiltonians we consider in this chapter. $\hat{Z}\subset Z^d$ is a subgroup of finite index $N=|Z^d/\hat{Z}|<s$ and each $\psi_q \in g(\mathscr{H}_0)$ is also $\hat{Z}$-periodic.

Consider now a small perturbation $\mathscr{H}=\mathscr{H}_0+\tilde{\mathscr{H}}$, $\tilde{\mathscr{H}}\in\boldsymbol{H}_s(\hat{Z})$ of $\mathscr{H}_0$, then the relative energy $\mathscr{H}(\Gamma_q)$ of contour $\Gamma^q\in\mathscr{C}_q$ decomposes as

$$\mathscr{H}(\Gamma^q)=\Psi(\Gamma^q)+\sum\nolimits_{m=1}^{r}[h(\psi_m)-h(\psi_q)]V_m(\Gamma^q), \tag{2.41}$$

where Ψ is a contour functional on $\mathscr{C}_q$ defined by (2.41) and h denotes specific energy with respect to $\mathscr{H}$. If $\tilde{\mathscr{H}}=0$, then $\Psi(\Gamma^q)=\mathscr{H}_0(\Gamma^q)$ and $\mathscr{H}_0(\Gamma^q)\geqq\varrho|\Gamma^q|$ holds with a positive constant ϱ in view of Peierls's stability condition. On the other hand (see (2.9)),

$$\left|\tilde{\mathscr{H}}(\Gamma^q)-\sum\nolimits_{m=1}^{r}[\tilde{h}(\psi_m)-\tilde{h}(\psi_q)]V_m(\Gamma^q)\right|\leqq 2(2s+1)^d\|\tilde{\mathscr{H}}\|\,|\Gamma^q|,$$

and $(2s+1)^d \leqq e^a$, consequently for each $\Gamma^q \in \mathscr{C}^q$, $q=1, 2, \ldots, r$, we have

$$\Psi(\Gamma^q) \geqq \frac{\varrho}{2} |\Gamma^q| \tag{2.42}$$

whenever $\|\tilde{\mathscr{H}}\| \leqq \frac{1}{4} e^{-a} \varrho$. Remember that $g(\mathscr{H}_0 + \tilde{\mathscr{H}}) \subset g(\mathscr{H}_0)$ follows from Lemma 2.2 in this case.

Proposition 2.6. *Let $\mathscr{H} = \mathscr{H}_0 + \tilde{\mathscr{H}}$, where the basic Hamiltonian $\mathscr{H}_0$ satisfies Peierls's stability condition with a constant $\varrho > 0$, while $\tilde{\mathscr{H}} \in \boldsymbol{H}_s(\hat{Z})$ and $\|\tilde{\mathscr{H}}\| \leqq \frac{1}{4} e^{-a} \varrho$. If $\beta \geqq (8a+1)/\varrho$, then there exists a unique vector $\hat{F} = (F_1, F_2, \ldots, F_r)$, $F_q = F_q(\Gamma^q)$, $\Gamma^q \in \mathscr{C}_q$, $q=1, 2, \ldots, r$, of $\hat{Z}$-periodic τ-functionals with $\tau = \frac{\beta\varrho - 1}{2} \geqq 4a$ such that*

$$\Theta(\Gamma^q | \beta\mathscr{H}) = \exp[b_q V(\Gamma^q)] \mathscr{Z}(\Gamma^q | F_q) \tag{2.43}$$

holds for each $\Gamma^q \in \mathscr{C}_q$, $q=1, 2, \ldots, r$, where

$$b_q = \beta h(\psi_q) - s(F_q) + \alpha \tag{2.44}$$

and α is defined by

$$\min_{1 \leqq q \leqq r} b_q = 0; \tag{2.45}$$

i.e. $b = (b_1, b_2, \ldots, b_r) \in 0_r$. Moreover, the unique solution $\hat{F} = \hat{F}(\beta\mathscr{H})$ is a Lipschitz-continuous function of $\beta\mathscr{H}$ (see (2.53)).

Proof. Observe that if $\mathscr{H}$ and β are given, then (2.43) is a system of equations for $\hat{F}$. For proving the existence of a unique solution $\hat{F}$, we rewrite (2.43) into the form

$$\hat{F} = \beta\hat{\Psi} + \hat{T}(\hat{F}_\mu, \beta\mathscr{H}),$$

where $\hat{\Psi} = (\Psi(\Gamma^1), \ldots, \Psi(\Gamma^r))$, and then we prove that the nonlinear operator $\hat{T}$ is a contraction in a suitably chosen complete metric space, at least if $\|\tilde{\mathscr{H}}\|$ and β^{-1} are small. Indeed, comparing (2.43) and Definition 2.12 we obtain a variant

$$\Theta_q(V | \beta\mathscr{H}) = \mathscr{Z}(V | F_q, b_q), \tag{2.43'}$$

of (2.43). Thus, substituting (2.43) and (2.43′) into (2.18),

$$\begin{aligned} \exp[b_q V(\Gamma^q)] \mathscr{Z}(\Gamma^q | F_q) &= \\ = \exp(-\beta\mathscr{H}(\Gamma^q)) \prod_{m=1}^{r} \mathscr{Z}(\mathrm{Int}_m \Gamma^q | F_m, b_m) \end{aligned} \tag{2.46}$$

follows for each finite $V \subset Z^d$ and $q=1, 2, \ldots, r$; (2.46) is again an equivalent form of (2.43). On the other hand, decomposing the logarithm of the right-hand side of

(2.20) according to (2.35), we have

$$\ln \mathscr{Z}(\Gamma^q|F_q) = s(F_q)V(\Gamma^q) - F_q(\Gamma^q) + \sum_{m=1}^r \Delta(\mathrm{Int}_m\, \Gamma^q|F_q),$$

while

$$\ln \mathscr{Z}(\mathrm{Int}_m\, \Gamma^q|F_m, b_m) = [s(F_m)+b_m]V_m(\Gamma^q) + \Delta(\mathrm{Int}_m\, \Gamma^q|F_m, b_m)$$

is the decomposition in Proposition 2.5. Therefore, taking the logarithm of both sides of (2.46) we obtain

$$[b_q+s(F_q)]\, V(\Gamma^q) - F_q(\Gamma^q) + \sum_{m=1}^r \Delta(\mathrm{Int}_m\, \Gamma^q|F_q) = \tag{2.47}$$

$$= -\beta\mathscr{H}(\Gamma^q) + \sum_{m=1}^r (s(F_m)+b_m)\, V_m(\Gamma^q) + \sum_{m=1}^r \Delta(\mathrm{Int}_m\, \Gamma^q|F_m, b_m),$$

whence by (2.41) and (2.44) it follows that

$$F_q(\Gamma^q) = \beta\Psi(\Gamma^q) + T_q(\hat{F}, \beta, \hat{h}), \tag{2.48}$$

where $\hat{h}=(h(\psi_1), \ldots, h(\psi_r))$ and

$$T_q(\hat{F}, \beta, \hat{h}) = \sum_{m=1}^r [\Delta(\mathrm{Int}_m\, \Gamma^q|F_q) - \Delta(\mathrm{Int}_m\, \Gamma^q|F_m, b_m)].$$

Although T_q depends explicitly on b, (2.44) and (2.45) determine its value as a function of $\beta\hat{h}$, so our notation is correct. Introducing the abbreviation $\hat{T}=(T_1, T_2, \ldots, T_r)$, (2.48) turns into

$$\hat{F} = \beta\hat{\Psi} + \hat{T}(\hat{F}, \beta, \hat{h}); \tag{2.43''}$$

(2.43″) is again an equivalent form of (2.43), and it follows easily from (2.43), (2.18) and (2.20) that any periodic solution $\hat{F}$ of (2.43″) is necessarily $\hat{Z}$-periodic.

Equation (2.43″) will be considered as an equation for $\hat{F}$ in the following complete metric space of vector-valued contour functionals $\hat{F}$. Let $\mathscr{B}_\tau$ denote the set of vectors $\hat{F}=(F_1, F_2, \ldots, F_r)$ such that $|\|\hat{F}\||=\max_{1\leqq q\leqq r} |\|F_q\||<\infty$ and each F_q is a $\hat{Z}$-periodic τ-functional on $\mathscr{C}_q$ with a common value of $\tau>0$; the distance of two points $\hat{F}'$ and $\hat{F}''$ in $\mathscr{B}_\tau$ is defined as $|\|\hat{F}'-\hat{F}''\||$. It is easy to check that $\mathscr{B}_\tau$ is indeed complete, while Proposition 2.3 implies that $\hat{T}$ is defined on $\hat{\mathscr{B}}_\tau$ if $\tau\geqq 4a$. First we prove that $\beta\hat{\Psi}+\hat{T}$ maps $\hat{\mathscr{B}}_\tau$ into itself, at least if $\beta\geqq\dfrac{2\tau+1}{\varrho}$ and $\tau\geqq 4a$. Indeed, (2.40) and Proposition 2.3 imply that

$$T_q(F, \beta h) \geqq -2e^{-\tau} \sum_{m=1}^r |\partial(\mathrm{Int}_m\, \Gamma^q)| \geqq -2e^{-\tau}3^d\, |\Gamma^q|; \tag{2.49}$$

therefore

$$\beta\Psi(\Gamma^q) + T_q(\hat{F}, \beta\hat{h}) \geqq \left[\frac{1}{2}\beta\varrho - 2e^{-\tau}3^d\right] |\Gamma^q| \geqq \tau|\Gamma^q|. \tag{2.50}$$

On the other hand, let $\mathscr{H}=\mathscr{H}_0+\tilde{\mathscr{H}}'$ and suppose that $\tilde{\mathscr{H}}'$ and β' satisfy the very same conditions as $\tilde{\mathscr{H}}$ and β. Then from Propositions 2.4 and 2.5 it follows that

$$|T_q(F, \beta h)(\Gamma^q) - T_q(F', \beta' h')(\Gamma^q)| \leqq 2 \sum_{m=1}^{r} |b_m - b'_m| V_m(\Gamma^q) +$$

$$+ \left(\frac{2}{e^a - 1} + 2e^{-\tau}\right) \sum_{m=1}^{r} \exp\left[a \operatorname{diam}(\operatorname{Int}_m(\Gamma^q))\right] V_m(\Gamma^q) \||\hat{F}_m - \hat{F}'_m\|| \leqq$$

$$\leqq 2(|\beta\hat{h} - \beta'\hat{h}'| + e^{-\tau}\||\hat{F} - \hat{F}'\||) V(\Gamma^q) + \frac{1}{2}\||\hat{F} - \hat{F}'\|| V(\Gamma^q) e^{a\delta(\Gamma^q)},$$

consequently

$$\||\hat{T}(\hat{F}, \beta\hat{h}) - \hat{T}(\hat{F}', \beta'\hat{h}')\|| \leqq 2|\beta\hat{h} - \beta'\hat{h}'| + \frac{1}{2}\||\hat{F} - \hat{F}'\|| \tag{2.51}$$

holds whenever $\hat{F}, \hat{F}' \in \mathscr{B}_\tau$. Choosing $\beta = \beta'$ and $\mathscr{H} = \mathscr{H}'$ we obtain the contraction property

$$\||\hat{T}(\hat{F}, \beta\hat{h}) - \hat{T}(\hat{F}', \beta\hat{h})\|| \leqq \frac{1}{2}\||\hat{F} - \hat{F}'\||. \tag{2.52}$$

Therefore for each $\beta \leqq \frac{8a+1}{\varrho}$ and $\mathscr{H} = \mathscr{H}_0 + \tilde{\mathscr{H}}$ with $\tilde{\mathscr{H}} \in \boldsymbol{H}_s(\hat{Z}), \|\tilde{\mathscr{H}}\| \leqq \frac{\varrho}{4} e^{-a}$ there exists a unique solution $\hat{F} = \hat{F}(\beta\mathscr{H})$ of (2.43) such that each component $\hat{F}_q$ of $\hat{F}$ is a τ-functional on $\mathscr{C}_q$ with $\tau = \frac{\beta\varrho - 1}{2} \geqq 4a$. Further, let Ψ' denote the contour functional corresponding to $\mathscr{H}' = \mathscr{H}_0 + \tilde{\mathscr{H}}'$. Then

$$|\beta\Psi(\Gamma^q) - \beta'\Psi'(\Gamma^q)| \leqq 2e^{a\delta(\Gamma)} \|\beta\mathscr{H} - \beta'\mathscr{H}'\| \, |\Gamma^q|$$

holds, i.e.

$$\||\beta\Psi - \beta'\Psi'\|| \leqq 2\|\beta\mathscr{H} - \beta'\mathscr{H}'\|;$$

while the comparison of (2.43″) and (2.52) results in

$$\frac{1}{2}\||\hat{F}(\beta\mathscr{H}) - \hat{F}(\beta'\mathscr{H}')\|| \leqq \||\beta\hat{\Psi} - \beta'\hat{\Psi}'\|| + 2|\beta\hat{h} - \beta'\hat{h}'|,$$

where $|\beta\hat{h} - \beta'\hat{h}'| \leqq \|\mathscr{H} - \mathscr{H}'\|$, so

$$\||\hat{F}(\beta\mathscr{H}) - \hat{F}(\beta'\mathscr{H}')\|| \leqq 8\|\beta\mathscr{H} - \beta'\mathscr{H}'\|. \tag{2.53}$$

Q.e.d.

Now we are in a position to prove Main Theorem A which, in turn, implies Theorem B. Notations introduced in the proof of Proposition 2.6 will be used without any reference.

Consider a perturbation $\mathscr{H}_\mu = \mathscr{H}_0 \sum_{i=1}^{r-1} \mu_i \mathscr{H}_i$ of $\mathscr{H}_0$ such that $\mathscr{H}_i \in \boldsymbol{H}_s(\hat{Z})$ for each $i = 1, 2, \ldots, r-1$ and $\mathscr{H}_\mu$ completely splits the degeneracy of the ground state of $\mathscr{H}_0$. Let $\mu=(\mu_1, \mu_2, \ldots, \mu_r)$,

$$\varepsilon_0 = \frac{\varrho}{4} e^{-a}\left[\sum_{i=1}^{r-1} \|\mathscr{H}_i\|\right]^{-1},$$

and let h_μ and h_i denote specific energy with respect to $\mathscr{H}_\mu$ and $\mathscr{H}_i$, respectively. Observe first that substituting $F(\beta\mathscr{H}_\mu)$ into (2.44) we can express $b=(b_1, b_2, \ldots, b_r)$ as a function $b=I_\beta(\mu)$ of $\mu\in U_0=\{\mu \,|\, |\mu|\leqq\varepsilon\}$ for each $\beta\geqq(8a+1)/\varrho$. We have to show that if β is large enough, then I_β is a homeomorphism of U_0 into 0_r and the image $I_\beta U_0$ contains an entire neighbourhood V_0 of 0 in 0_r; let $V_0=\{b\in 0_r \,|\, |b|<1\}$. The continuity of I_β is a direct consequence of (2.53) and of Proposition 2.4. The inverse mapping to I_β can be described in the following way.

Suppose that $\mathscr{H}_\mu$ is fixed, then

$$h_\mu(\psi_q)-h_\mu(\psi_{q+1}) = \sum_{i=1}^{r-1} \mu_i[h_i(\psi_q)-h_i(\psi_{q+1})], \quad q = 1, 2, \ldots, r-1$$

is a system of linear equations for μ. Since the perturbation $\mathscr{H}_\mu$ completely splits the degeneracy of the ground state of $\mathscr{H}_0$, for each q there exists a $\mu=\mu(q)$ such that $\mathscr{g}(\mathscr{H}_\mu)=\mathscr{g}(\mathscr{H}_0)-\{\psi_q\}$; i.e. the rank of the $(r-1)\times(r-1)$ matrix $(h_{qi})=(h_i(\psi_q)-h_i(\psi_{q+1}))$ is exactly $r-1$. This means that each μ_i can be expressed as a linear combination of the differences

$$h_\mu(\psi_q)-h_\mu(\psi_{q+1}) = \frac{1}{\beta}[b_q+s(F_q)-b_{q+1}-s(F_{q+1})];$$

i.e.

$$\mu_i = \frac{1}{\beta} \sum_{q=1}^{r} a_{iq}(b_q+s(F_q)), \quad i = 1, 2, \ldots, r-1 \tag{2.54}$$

holds with some coefficients a_{iq} depending only on the values $h_j(\psi_m)$, $1\leqq j<r$, $1\leqq m\leqq r$.

Since $F_q=F_q(\beta\mathscr{H}_\mu)$, (2.54) can be considered as a nonlinear system of equations for finding μ if β and b are given. Let

$$L = \max_{1\leqq i<r} \sum_{q=1}^{r} |a_{iq}|$$

and suppose that $\beta\geqq(8a+1)/\varrho$. Then Proposition 2.3 implies

$$\max_{1\leqq i<r}\left|\frac{1}{\beta} \sum_{q=1}^{r} a_{iq}\left[b_q+s(F_q(\beta\mathscr{H}_\mu))\right]\right| \leqq \frac{2L}{\beta}$$

for each $b\in V_0$ and $\mu\in U_0$, while

$$\max_{1\leqq i<r}\left|\frac{1}{\beta} \sum_{q=1}^{r} a_{iq}\, s(F_q(\beta\mathscr{H}_\mu))-\frac{1}{\beta} \sum_{q=1}^{r} a_{iq}\, s(F_q(\beta\mathscr{H}_{\mu'}))\right| \leqq$$

$$\leqq \frac{L}{\beta}|\mu-\mu'|e^{-(\beta\varrho-1)/2}$$

follows from Proposition 2.4 whenever $\mu, \mu' \in U_0$. Therefore if $\beta \geqq \beta_0$ and $b \in V_0$, where β_0 is so large that $\beta_0 \geqq \frac{8a+1}{\varrho}$, $\frac{2L}{\beta_0} \leqq \varepsilon_0$ and $\frac{L}{\beta_0} \exp\left[-\frac{\beta\varrho-1}{2}\right] < 1$, then the right-hand side of (2.54) is a contraction of U_0 into itself. This proves that I_β is a homeomorphism of U_0 into 0_r and $I_\beta U_0 \supset V_0$.

Consider now a component F_q of $\hat{F} = \hat{F}(\beta\mathscr{H}_\mu)$ such that $b_q = 0$. Let $P_V(\cdot|F_q)$ denote the contour model defined by F_q in $V \subset Z^d$, while $P_V^0(\cdot|\psi_q, \beta\mathscr{H}_\mu)$ denotes the conditional Gibbs distribution in $V \subset Z^d$ given $\varphi(\bar{V}) = \psi_q(\bar{V})$ and $(\partial\varphi, \varphi(\partial p)) \subset V$; P_V^0 is just the conditional Gibbs distribution corresponding to the dilute statistical sum $\Theta_q(V|\beta\mathscr{H}_\mu)$. Since $\mathscr{Z}(\Gamma^q|F_q) = \Theta(\Gamma^q|\beta\mathscr{H}_\mu)$ and $\mathscr{Z}(V|F_q) = \Theta_1(V|\beta\mathscr{H}_q)$, the joint distribution of external contours is the same in both cases, so the results of Section 8 can be reformulated for lattice systems. Namely, if $V \to \infty$, then there exists the weak limit of the distribution of external contours in the lattice systems P_V^0 and the limit distribution is the very same as described in Proposition 2.2. On the other hand, given the set θ of external contours, the configurations $\varphi(\operatorname{Int}\Gamma)$, $\Gamma \in \theta$, inside the external contours are conditionally independent with distributions $P_{\operatorname{Int}\Gamma}^0(\cdot|\psi_q, \beta\mathscr{H}_\mu)$; thus the model of external contours provides a complete description of the lattice system. Hence it follows easily that the limit

$$P_{\beta,\mu}^{(q)} = \lim_{V\to\infty} P_V^0(\cdot|\psi_q, \beta\mathscr{H}_\mu) = \lim_{V\to\infty} P_V(\cdot|\psi_q, \beta\mathscr{H}_\mu)$$

exists, and that $P_{\beta,\mu}^{(q)}$ describes small local distortions of ψ_q as $\beta \to \infty$. Moreover, the exponential mixing property of external contours implies that $P_{\beta,\mu}^{(q)}$ is ergodic, and hence it is an indecomposable limit Gibbs distribution; i.e. $P_{\beta,\mu}^{(q)}$ is really a pure phase. Q.e.d.

Proposition 2.6 is of fundamental importance in the proof of the main result, because it reveals the correspondence existing between lattice systems and contour models with or without a parameter. Namely, the statistical sums $\Theta(\Gamma^q|\beta\mathscr{H}_\mu)$ of the lattice system coincide either with the crystal statistical sum of a contour model or with that of a contour model with a parameter. The recognition of this somewhat unexpected fact has opened the way towards proving the basic theorems. On the other hand, Proposition 2.6 contains the condition of coexistence of phases. Indeed, if Γ^q is an external contour in a pure phase $P_{\beta,\mu}^{(q)}$ at $\mu = \bar{\mu}(\beta)$, i.e. $b_q = 0$ for each $q = 1, 2, \ldots$ $\ldots, r$, then it is quite natural to claim that $\operatorname{Int}_m \Gamma^q$ is occupied by the mth phase and $\operatorname{supp}\Gamma^q$ is a surface separating coexisting phases. In view of Peierls's inequality for the model of external contours (see Proposition 2.2), we have

$$\tilde{\varrho}(\Gamma^q) \leqq \exp\left[-\frac{1}{2}(\beta\varrho - 1)|\Gamma^q|\right]$$

for the probability $\tilde{\varrho}(\Gamma^q)$ of the event that Γ^q is an external contour. On the other hand, (2.33) implies that

$$\tilde{\varrho}(\Gamma^q) \geqq e^{-F_q(\Gamma^q)}\left(1-\sum_{n=9}^{\infty} \exp\left[an-\frac{n}{2}(\beta\varrho-1)\right]\right) \geqq$$

$$\geqq \left(1-\exp\left[-\frac{1}{2}(\beta\varrho-1)\right]\right)\exp\left[-\frac{1}{2}\|F_q\|(\beta\varrho-1)|\Gamma^q|\right].$$

These relations result in the following condition of coexistence of phases. If we replace a phase by an other one in a certain volume, then the gain of free energy is only proportional to the separating boundary rather than to the volume!

The above considerations apply only to phases described by contour models. The importance of parametric contour models lies in describing thermodynamically unstable phases (see Martirosyan [23, 24]). His main result is the so-called theorem on the strip stating that in the case of such unstable boundary conditions only a thin strip at the boundary is occupied by the unstable phase, other domains in a typical configuration correspond to stable phases; i.e. a boundary condition ψ_q with $b_q > 0$ does not give rise to a long-range order. It would be very interesting to continue these investigations towards proving in full generality that there are no indecomposable and periodic limit Gibbs distributions other than those described in the Main Theorem.

11. Some further remarks

1. Our definition of ground states (see Section 2) does not presuppose that they are periodic configurations; of course, there may also exist non-periodic ground states. For example, let $\mathscr{H}_0$ denote the Hamiltonian of the d-dimensional Ising ferromagnet and define $\psi_{a,i} \in \Omega$ by $\psi_{a,i}(x)=1$ if $x=(x_1, \ldots, x_d)$ and $x_i < a$, $\psi_{a,i}(x)=-1$ otherwise. It is easy to check that each $\psi_{a,i}$, $a \in Z^1$, $i=1, 2, \ldots, d$, is a ground state of $\mathscr{H}_0$. However, the existence problem of limit Gibbs states describing local distortions of such non-periodic ground states is far from being trivial. Gallavotti and Miracle-Sole [68] and Messager and Miracle-Sole [102] have shown that in the two-dimensional case there are no limit Gibbs distributions associated with such non-periodic ground states. The intuitive background of this phenomenon is the very same as that of the absence of phase transitions in one-dimensional systems. Namely, the relative energy $\mathscr{H}(\varphi|\psi)$ with respect to such a ground state ψ may remain bounded even if $\varphi \neq \psi$ in large domains. In other words, such ground states are unstable in a very natural sense. However, if $d > 2$, then the situation is quite different. Dobrushin [17] has shown that if $d > 2$ and β is large enough, then limit Gibbs distributions do exist describing small local distortions of non-periodic ground states (see also Remark 6 in the next section). It would be very interesting to have a general theory

describing conditions for the coexistence of pure phases associated with non-periodic ground states at large β.

2. Recently Glimm, Jaffe and Spencer [77, 78] proved the existence of a phase transition in the $\lambda{:}\varphi^4{:}_2$ euclidean quantum field theory for large values of λ/m_0^2, where m_0 denotes the bare mass. Following [106] we introduce here the concept of ground states for the lattice approximation of $\lambda{:}p(\varphi){:}_2$ quantum field theory, and show that at large values of λ/m_0^2 the principal background phenomenon appears of a phase transition that is a degenerate ground state.

For simplicity we consider a system $\varphi(x)$, $x\in Z_h^2$, of continuous spins on the two-dimensional lattice Z_h^2 of lattice size h in the limit when h goes to zero. The spins $\varphi(x)$ are unbounded real variables and the formal Hamiltonian $\mathscr{H}$ of the system under consideration is of the form

$$\mathscr{H} = \mathscr{H}_0 + \sum_{x\in Z_h^2} \lambda h^2 :P(\varphi(x)):_{\sigma(m_0)},$$

where

$$\mathscr{H}_0 = \sum_{x\in Z_h^2} m_0^2 h^2 :\varphi^2(x):_{\sigma(m_0)} + \frac{1}{2}\sum_{x\in Z_h^2}\sum_{\|x-y\|=h} (\varphi(x)-\varphi(y))^2$$

is the free-field Hamiltonian,

$$:P(\varphi):_{\sigma(m_0)} = \sum_{k=1}^{2n} p_k :\varphi^k:_{\sigma(m_0)}$$

and $:\varphi^k:_{\sigma(m_0)}$ is the kth Hermite polynomial with respect to the centered Gaussian distribution of variance $\sigma(m_0)$; $\sigma(m_0)$ is just the variance of an individual spin variable in the free field. It is easy to check that $\sigma(m_0)\sim \text{const}\cdot\ln\frac{1}{h}$ as $h\to 0$ and $\sigma(m_0)-$ $-\sigma(m_1)\sim\text{const}\cdot\ln\frac{m_1}{m_0}$ uniformly in h as $m_1\to\infty$.

If one tries to define periodic ground states of $\mathscr{H}$ as periodic configurations of minimal specific energy, it turns out that such configurations tend to $\pm\infty$ as $h\to 0$. Therefore such a definition is not satisfactory because it presupposes that fluctuations of the field around the ground state are uniformly small. Fluctuations, however, should be small only in the mean when one considers averages of the field over small but macroscopic domains.

The correct definition of ground states is as follows. For simplicity we consider only constant ground states, i.e. $\varphi(x)=A$ for each x. Substitute $\varphi=A+\psi$ into $\mathscr{H}$ and suppose that we obtain

$$\mathscr{H} = \mathscr{H}(\psi) = \frac{1}{2}\sum_x\sum_{\|x-y\|=h}(\psi(x)-\psi(y))^2 + \tag{2.55}$$
$$+\sum\left[h^2\left(m^2:\psi^2:_{\sigma(m)}+\sum_{k=3}^{2n} q_k(A):\psi^k:_{\sigma(m)}\right)+U(A)\right],$$

$U(A)$ and $m=m(A)>0$ are constants, $q_3, \ldots, q_{2n}$ are some coefficients. Expression (2.55) shows that the Hamiltonian $\mathscr{H}(\psi)$ has a free-field part of bare mass $m=m(A)$, while the polynomial $\sum q_k(A) :\psi^k:_{\sigma(m)}$ can be considered as an interaction part; the equation expresses the fact that A is a state of local minimum and $U=U(A)$ should be considered as the value of the depth of this local minimum. Of course, A and m do depend on the dimensionless parameter λ/m_0^2 of the theory.

Suppose that (2.55) holds with some functions

$$A = A\left(\frac{\lambda}{m_0^2}\right) \quad \text{and} \quad m = m\left(\frac{\lambda}{m_0^2}\right), \quad \text{and that further} \quad \frac{q_k}{m} \to 0 \quad \text{if} \quad \frac{\lambda}{m_0^2} \to \infty$$

for each $k=3, \ldots, 2n$. Then the pair $A\left(\frac{\lambda}{m_0^2}\right)$, $m\left(\frac{\lambda}{m_0^2}\right)$ is called an asymptotic local ground state (ALGS) of bare mass $m=m\left(\frac{\lambda}{m_0^2}\right)$.

If A is an ALGS, then the interaction part of (2.55) turns out to be a small perturbation of the free-field part. Thus it is quite natural to expect that translation-invariant vacuum states for $\mathscr{H}$ describe small fluctuations near the configuration $\varphi\equiv A$ in such a way that locally typical realizations look like $\psi+A$, where ψ is a typical realization of the free field of bare mass m. As we shall see, $m\left(\frac{\lambda}{m_0^2}\right)\to\infty$ when $\frac{\lambda}{m_0^2}\to\infty$, and this shows in what sense these fluctuations are small. It will be shown that the set of ALGS is finite.

An ALGS $\left(A\left(\frac{\lambda}{m_0^2}\right), m\left(\frac{\lambda}{m_0^2}\right)\right)$ is called an absolute ALGS if $U(A)=\min U(A')$ where the minimum is taken over all ALGS. Bearing in mind that $U(A)$ is just the first term in the expansion of the logarithm of the statistical sum when the interaction part of the Hamiltonian is considered to be small, it is natural to expect that only absolute ALGS can give rise to different vacuum states.

Now we proceed to the derivation of some equations for A and m. Since $:\varphi^k:_{\sigma(m_0)}= =:(:\varphi^k:_E):_{\sigma(m_0)}$ if $E=\sigma(m_0)-\sigma(m)$, we have

$$\mathscr{H}(\psi) = \frac{1}{2}\sum_{\|x-y\|=h}(\psi(x)-\psi(y))+$$

$$+\sum_x h^2\left[m_0^2:(\psi+A)^2:_{\sigma(m_0)}+\lambda\sum_{k=1}^{2n} p_k:(\psi+A)^k:_{\sigma(m_0)}\right]=$$

$$=\frac{1}{2}\sum_{\|x-y\|=h}(\psi(x)-\psi(y))^2+\sum_x h^2\left(m_0^2:(\psi+A)^2:_E+\lambda\sum_{k=1}^{2n} p_k:(\psi+a)^k:_E\right)_{\sigma(m)}$$

and it is easy to check that for fixed m_0 and m the limit of E exists as $h\to 0$ and its value is proportional to $\ln\frac{m}{m_0}$. Hence it follows that

$$\lambda\sum_{k=0}^{2n} p_k H_E^{(k)}(A) = U(A), \tag{2.56}$$

(2.57) $$2m_0^2 A+\lambda \sum_{k=1}^{2n} \frac{dH_E^{(k)}(A)}{dA}=0,$$

(2.58) $$2m_0^2+\lambda \sum_{k=1}^{2n} \frac{d^2 H_E^{(k)}(A)}{dA^2}=m^2,$$

where $H_E^{(k)}(\psi)=:\psi^k:_E$. Here (2.57) and (2.58) determine A and m^2, while (2.56) is the expression for the depth of the minimum. Remember that, in the limit $h\to 0$, a model of quantum field theory can be obtained; after this we can let $\frac{\lambda}{m_0^2}\to\infty$. Therefore it is very important that all quantities in the equations above have a well-defined limit as $h\to 0$.

Now we turn to the study of the asymptotic behaviour of solutions of (2.56)–(2.58) when $\frac{\lambda}{m_0^2}\to\infty$. Let $M=\frac{m^2}{m_0^2}$. We expect that $M\to\infty$ and $E\to\infty$ as $\frac{\lambda}{m_0^2}\to\infty$. Indeed, as $H_E^{(k)}(t)=E^{n/2}H_1^{(k)}\left(\frac{t}{E^{1/2}}\right)$, (2.58) can be rewritten as

$$2+E^{n-1}\frac{\lambda}{m_0^2}\left[\frac{d^2}{dt^2}H_1^{(2n)}(B)+\ldots\right]=2M^2,$$

where $B=A/E^{1/2}$ and the remainder terms indicated only by dots are negligible if E is large. Hence we obtain the asymptotic relation

(2.59) $$\frac{\lambda}{m_0^2}\left[\frac{d^2}{dt^2}H_1^{(2n)}(B)\right]\approx\frac{2M^2}{E^{n-1}},$$

and the right-hand side becomes proportional to $M^2/(\ln M^2)^{n-1}$ as $h\to 0$. On the other hand, (2.57) turns into

(2.60) $$\frac{d}{dt}H_1^{(2n)}(B)\approx 0,$$

it is an equation for A provided that m and E are given; let b_i, $i=1,2,\ldots,2n-1$, denote the roots of the equation $\frac{d}{dt}H_1^{(2n)}(B)=0$. Using the usual perturbation theory it is easy to show that b_i is an approximate root of (2.60) if $E\to\infty$, provided that $\frac{d^2}{dt^2}H_1^{(2n)}(B_i)\neq 0$. We may assume that $\frac{d^2}{dt^2}H_1^{(2n)}(b_i)>0$ because this case corresponds to a local minimum of H. Thus the ground states are $A_i\approx E^{1/2}b_i$, while from (2.59) we obtain that $M_i^2\approx\text{const}\cdot\frac{\lambda}{m_0^2}\left[\ln\frac{\lambda}{m_0^2}\right]^{n-1}$ and $q_k(A_i)\approx\text{const}\cdot\lambda\left[\ln\frac{\lambda}{m_0^2}\right]^{n-ik/2}$ if $k=3,4,\ldots,2n$. Consequently, $m_i^{-2}q_k(A_i)\to 0$ when $\frac{\lambda}{m_0^2}\to\infty$ and $(E^{1/2}B_i, m_i)$ is an asymptotic local ground state. For $U(A_i)$ we get

$$\frac{U(A_{i_1})}{U(A_{i_2})} \approx \frac{H_1^{(2n)}(B_{i_2})}{H_1^{(2n)}(B_{i_1})},$$

i.e. the absolute minima of $U(A_i)$ correspond to those of $H_1^{(2n)}(t)$. Hermite polynomials $H_1^{(2n)}(t)$ for $n \leqq 5$ have exactly two absolute minima; it is not known whether this is true or not for all n. But in general it is natural to expect that for our Hamiltonian $\mathscr{H}$ the number of translation-invariant vacuum states at large values of $\frac{\lambda}{m_0^2}$ is not more than the number of absolute minima of $H_1^{(2n)}$.

3. We suggest a new formulation of the Main Theorem. Consider a real function $y=f(x)$ having r minima $b_1, b_2, \ldots, b_r$, i.e. $b_i \neq b_j$ if $i \neq j$, $f(b_i)=f(b_j)=\bar{f}$ and $f(x)>\bar{f}$ if $x \notin \{b_1, b_2, \ldots, b_r\}$. Such a function has a rather particular structure; in view of the general theory of bifurcations (see [1]) it is possible to construct a versal family $y=f(x, \lambda)$, $\lambda=(\lambda_1, \lambda_2, \ldots, \lambda_r)$ such that $f(x)=f(x, 0)$ and the parameter space has a stratification. This means that the parameter space splits into the point $\lambda=0$, r curves, C_r^2 two-dimensional surfaces and so on, in such a way that on each of the C_r^k k-dimensional surfaces f has exactly k equal minima. Such a stratification is universal in a well-defined sense. Although our formal Hamiltonian $\mathscr{H}_0(\varphi)$ cannot be considered as a real function and its stable ground states cannot be interpreted as points of isolated minima, the phase diagram has the very same structure as a stratification given by a versal family as above. It would be interesting to study in more detail the smoothness properties of the strata of the phase diagram and to show that

$$\mathscr{H}_0 + \mu_1 \mathscr{H}_1 + \ldots + \mu_{r-1} \mathscr{H}_{r-1}$$

is a versal family for $\mathscr{H}_0$ in a well-defined sense.

Historical notes and references to Chapter II

1. Uniqueness of the limit Gibbs distribution for a Hamiltonian $\beta\mathscr{H}$ at large values of β is proved in the book by Ruelle [39] and in the papers of Dobrushin [13–16].

2. The Ising ferromagnet in dimensions $d \geqq 2$ and at large values of β was investigated by Peierls [105] in 1936. The fundamental inequality for the probability of contours (see Section 3) appeared also there. The Ising model is discussed in the same spirit in the papers by Griffiths [79] and by Dobrushin [12].

3. An analysis of limit Gibbs distributions based on the study of the statistics of contours is often referred to as Peierls's method of contours. This method has been very useful in the investigation of several classes of lattice systems, for example Berezin and Sinai [2], Ginibre, Grossman and Ruelle [72], Dobrushin [55], Gercik and Dobrushin [10], Gercik [11], Heilmann and Lieb [84], Heilmann [85], Pirogov [33, 34], Pirogov and Sinai [35], Cassandro, Fano and Olivieri [51], Runnels [111]. Some interesting results have been obtained by Slawny [114] and Holsztynski and Slawny [86]. The general theory discussed in this chapter is contained in the papers by Pirogov and Sinai [36, 37].

Bortz and Griffiths [49], Malyshev [100] and Lieb [98] prove the coexistence of phases in some lattice systems with continuous spin $\varphi(x)$. It would be very interesting to extend the general theorems of this chapter to continuous lattice systems.

In the book by Gruber, Hintermann and Merlini [80] certain algebraic concepts of duality are used to derive some interesting results.

4. The ground states of some lattice systems have been described by Kashapov [21]. Holsztynski and Slawny [87] prove several general results on the structure of the set of ground states.

Contour models have been introduced by Minlos and Sinai [27–29], where the cluster properties and also some further questions on correlation functions are discussed, too.

5. Gallavotti and Miracle-Sole [68] proved that in the Ising ferromagnet at large values of β there are only two translation-invariant limit Gibbs states. In the two-dimensional case, Messager and Miracle-Sole have extended this result to all values of β above the critical one.

Martirosyan has shown that for such values of the parameters when the existence of r or $(r-1)$ pure phases is stated in the Main Theorem, there are no other periodic limit Gibbs states. (See [23], the detailed proof of these results is to appear in *Izvestia Acad. Sci. Armenia, Ser. Math.*) Some other related results can be found in the paper by Martirosyan [24].

6. Dobrushin [17] proved that at large values of β some non-periodic limit Gibbs states associated with non-periodic ground states also exist. In case of the three-dimensional Ising ferromagnet a very elegant simple proof of the above result has been given by van Beijeren [46].

7. A very general approach to the construction of Hamiltonians with several associated limit Gibbs states has been developed by Ruelle [40, 110]; the method is based on the fairly general results by Israel [89]. The connections between this approach and that of this chapter are not quite clear yet.

8. The Main Theorem of this chapter can be interpreted as a variant of the rigorous proof of the Gibbs phase rule (cf. Lebowitz [96]).

9. The coexistence of phases in two-dimensional quantum fields has been proved by Glimm, Jaffe and Spencer [77, 78]. Our description of the structure of the set of ground states follows Pirogov and Sinai [106]. Some recent results by Gawedzki [71] are also interesting.

10. Recently Calogero has pointed out (private communication) that the answer to the question formulated at the end of Remark 2, Section 11 is positive, i.e. each even Hermite polynomial has exactly two absolute minima (see Section 10.18 in [118, Vol. 2]).

CHAPTER III

LATTICE SYSTEMS WITH CONTINUOUS SYMMETRY

1. Introduction

In this chapter we consider classical lattice systems with continuous symmetry. The space Φ of spin variables $\varphi(s)$ will be a homogeneous space, acted on by a Lie group G. The main example is $\Phi=S^\nu$, the ν-dimensional sphere, with $G=SO(\nu+1)$, the group of the $(\nu+1)$-dimensional orthogonal matrices with determinant 1. The interaction is assumed to be translation-invariant, with finite radius. The Hamiltonian of such a system is given by a potential $\mathscr{U}(\varphi(W_R(s))$, where $W_R(s)$ is a ball of radius R and centre s, $\varphi(W_R(s))$ being the configuration $\{\varphi(t), t\in W_R(s)\}$. The potential $\mathscr{U}(\varphi(W_R(s))$ describes the interaction of the variable $\varphi(s)$ with its neighbours in the ball $W_R(s)$. The potential $\mathscr{U}(\varphi(W_R(s))$ is said to be invariant with respect to the group G if $\mathscr{U}(\varphi(W_R(s))=\mathscr{U}(g\varphi(W_R(s))$ for any $g\in G$, where $g\varphi(W_R(s))=\{g\varphi(t), t\in W_R(s)\}$. The Hamiltonian $\mathscr{H}$ is formally given as

$$\mathscr{H}(\varphi)=\sum_{s\in Z^d}\mathscr{U}(\varphi(W_R(s)). \tag{3.1}$$

If the potential $\mathscr{U}$ is invariant with respect to G, the Hamiltonian $\mathscr{H}$ is also called invariant with respect to G. Let us assume that a normed measure μ is given on G, which is invariant under the action of the group G.

The conditional Gibbs distributions in the space of configurations φ_V under the condition $\varphi(Z^d-V)$ and the limit Gibbs distributions corresponding to them can be constructed by the help of the Hamiltonian (3.1) and measure μ in the usual way.

Just as in the discrete case, investigated in detail in Chapter II, the structure of the limit Gibbs distributions corresponding to the Hamiltonian $\beta\mathscr{H}$ for large β is closely connected to the ground states of $\mathscr{H}$. However, this connection is not as simple as in the discrete case. The notion of stability of a ground state turns out to be more delicate than Peierls's condition.

Similarly to the discrete case, a configuration $\psi=\{\psi(s), s\in Z^d\}$ is called a ground state of the potential $\mathscr{H}$, if $\mathscr{H}(\psi|\psi)=\mathscr{H}(\varphi)-\mathscr{H}(\psi)\geqq 0$ for any configuration φ agreeing with ψ almost everywhere. If the Hamiltonian $\mathscr{H}$ is invariant with respect to G, and ψ is a ground state for $\mathscr{H}$, then $g\psi$ is also a ground state for an arbitrary

$g \in G$. Thus the set of the ground states is a G-space, i.e. a space where the group G acts.

Let us consider the following special case: $\mathscr{U}(\varphi(W_R(s)))$ is a C^∞ function on $\prod_{t \in W_R(s)} \Phi(t)$, and $\mathscr{U}(\varphi(W_R(s)))=0$, if $\varphi(t_1)=\varphi(t_2)$ for every $t_1, t_2 \in W_R(s)$, and $\mathscr{U}(\varphi(W_R(s)))>0$ otherwise. In other words, $\mathscr{U}$ takes its minimum on the diagonal $\mathscr{D} \subset \prod_{t \in W_R(s)} \Phi(t)$, $\mathscr{D}=\{\varphi(W_R(s)):\varphi(t) \equiv \text{const}\}$. It is clear that, in this case, every ground state ψ has the form $\psi=\{\varphi(s)=\text{const}\}$, and the set of ground states ψ is isomorphic to Φ in a natural way.

Let us now consider the question about the stability of the ground states. Let $\bar{\varphi} \in \Phi$ and $\psi_{\bar{\varphi}}=\{\varphi(s) \equiv \bar{\varphi}\}$. Let $\mathscr{T}_{\bar{\varphi}}$ denote the tangent space to Φ at the point $\bar{\varphi}$, and let us consider the formal infinite sequence of tangent vectors $\tau=\{\tau(s) \in \mathscr{T}_{\bar{\varphi}}, s \in Z^d\}$. Let us write again a formal small perturbation of the ground state $\psi_{\bar{\varphi}}$ in the form $\tilde{\psi}=\{\bar{\varphi}+\varepsilon\tau(s), s \in Z^d\}$, where ε is a small parameter. Then $\bar{\varphi}+\varepsilon\tau(s)$ can be considered as a point of Φ, obtained, e.g. by the exponential mapping. In the Hamiltonian every term $\mathscr{U}(\varphi(W_R(s)))$ can be differentiated with respect to ε at $\varepsilon=0$, and we can write

$$(3.2) \quad \mathscr{H}(\psi) = \mathscr{H}(\psi_{\bar{\varphi}})+\frac{\varepsilon^2}{2} \sum_{s \in Z^d} \sum_{t_1, t_2} \left(\frac{\partial^2 \mathscr{U}(\varphi(W_R(s)))}{\partial\varphi(t_1)\partial\varphi(t_2)} \bigg|_{\varphi(t)=\bar{\varphi}} \tau(t_1), \tau(t_2) \right) + O(\varepsilon^2).$$

The linear terms in ε vanish since $\psi_{\bar{\varphi}}$ is a ground state. The expression $\dfrac{\partial^2 \mathscr{U}(\varphi(W_R(s)))}{\partial\varphi(t_1)\partial\varphi(t_2)}$ must be considered here as an operator on the conjugate space $\mathscr{T}^*_{\bar{\varphi}}$. If $\tau(t) \equiv \tau$, then

$$\sum_{t_1} \frac{\partial^2 \mathscr{U}(\varphi(W_R(s)))}{\partial\varphi(\tau_1)\partial\varphi(\tau_2)} \tau = 0$$

since the whole diagonal $\mathscr{D}$ consists of the minima of the function $\mathscr{U}(\varphi(W_R(s)))$. Therefore the quadratic form in (3.2) can be written in the form

$$\mathscr{H}_1(\tau) = \sum_{s \in Z^d} \sum_{t_1, t_2} \left(\frac{\partial^2 \mathscr{U}(\varphi(W_R(s)))}{\partial\varphi(t_1)\partial\varphi(t_2)} (\tau(t_1)-\tau(t_2)), (\tau(t_1)-\tau(t_2)) \right).$$

Let us consider a Gaussian distribution on the space of sequences $\tau=\{\vec{\tau}(s), s \in Z^d\}$ with the Hamiltonian $\mathscr{H}_1(\tau)$ (see Chapter I, Section 4, Point 3). Such a probability distribution may be a generalized one, i.e. the expectation $\langle\tau(s), \tau(s)\rangle$ determined by this distribution may be infinite. This means that only the expectations of the type $\langle\tau(s_1)-\tau(s_2), \tau(s_1)-\tau(s_2)\rangle$ or those with higher order difference are finite, i.e. τ generates a process with finite-order stationary increments. In this case, we call the ground state unstable. The reason for this terminology is that a small change of energy (of the order of $\varepsilon^2\mathscr{H}_1$) corresponds to large values of τ, i.e. the fluctuation around $\psi_{\bar{\varphi}}$ is large. On the other hand, if the expectation $\langle\tau(s), \tau(s)\rangle$ is finite, then the ground state is called stable.

Example. The classical Heisenberg model

In this example $\Phi = S^{d-1}$, the interaction radius $R=1$, and

$$\mathscr{U}(\varphi(W_1(s_1))) = 1 - \frac{1}{2d} \sum_{\|s_1 - s_2\| = 1} (\varphi(s_1), \varphi(s_2)).$$

Furthermore

$$\mathscr{H}(\tilde{\psi}) = \mathscr{H}(\psi_{\bar{\varphi}} + \varepsilon\tau) =$$

$$= \sum_{s \in Z^d} \left[1 - \frac{1}{2d} \sum_{\|s - s_2\| = 1} \frac{(\bar{\varphi} + \varepsilon\tau(s), \bar{\varphi} + \varepsilon\tau(s_2))}{\sqrt{1 + \varepsilon^2 \|\tau(s)\|^2} \sqrt{1 + \varepsilon^2 \|\tau(s_2)\|^2}} \right] \approx$$

$$\approx \mathscr{H}(\psi_{\bar{\varphi}}) + \frac{\varepsilon^2}{2} \sum_{\|s_1 - s_2\| = 1} (\tau(s_1) - \tau(s_2), \tau(s_1) - \tau(s_2)).$$

It is not difficult to see that every ground state $\psi_{\bar{\varphi}}$ is unstable for $d=2$, and is stable for $d \geqq 3$.

The main hypothesis about lattice systems with continuous symmetry, considered in this chapter, is that at large β there is no limit Gibbs distribution arising from a small perturbation of an unstable ground state, and there exists such a limit Gibbs distribution if the ground state is stable.

In this chapter, two results supporting the above hypothesis are proved. In Section 2 a theorem of Dobrushin and Shlosman is proved stating that in two-dimensional systems, under very general conditions, every limit Gibbs distribution is invariant under the action of the symmetry group G of the Hamiltonian. This result can be considered as a sharpening of the well-known Mermin–Wagner inequality, which implies the invariance of the pair correlation function under G. In Section 3, a theorem of Fröhlich, Simon and Spencer is proved. It asserts that, in the classical Heisenberg model, for $d \geqq 3$ there is a long-range order if β is large, which implies the existence of limit Gibbs distributions corresponding to a small perturbation of the ground state $\psi_{\bar{\varphi}}$.

2. Absence of breakdown of continuous symmetry in two-dimensional models

Let us first consider the case when the group G is the unit circle with the usual multiplication. Let G act on a measurable space $(\Phi, \mathscr{F})$ where a G-invariant measure μ is given. This means that, for any $F \in \mathscr{F}$, $g \in G$, $\mu(gF) = \mu(F)$. We consider a two-dimensional model with a translation invariant potential $\mathscr{U}(\varphi(W_R(s)))$ of finite radius R. Given a point s, two points $t_1, t_2 \in W_R(s)$, a configuration $\varphi(W_R(s))$ on $W_R(s)$ and $g_1, g_2 \in G$, $F(g_1, g_2, t_1, t_2, \varphi(W_R(s)))$ denotes the configuration on $W_R(s)$ which has the value $\varphi(t)$ at t if $t \neq t_1, t_2$ and the value $g_i \varphi(t_i)$, $i=1, 2$, at t_1 and t_2. Fixing s, t_1, t_2

and $\varphi(W_R(s))$, we define the function

$$k(g_1, g_2) = \mathcal{U}(F(g_1, g_2, t_1, t_2, \varphi(W_R(s))).$$

We assume that k is a twice differentiable function of its arguments, and

$$\text{A)} \qquad \max\left|\frac{\partial^2 k}{\partial g_1 \partial g_2}\right| \leqq L,$$

where the maximum is taken in $g_1, g_2, s_1, \varphi(W_R(s)), t_1$ and t_2; L is an appropriate constant.

Theorem 3.1. *(see Dobrushin and Shlosman [56]). Every limit Gibbs distribution corresponding to the Hamiltonian $\beta\mathcal{H}$, $0<\beta<\infty$, is invariant under the group G.*

Proof. It is sufficient to consider the case $\beta=1$. We introduce the squares V_n, $n=1, 2, \ldots$,

$$V_n = \{t = (t_1, t_2) \in Z^2, \ |t_1| \leqq n, \ |t_2| \leqq n\}.$$

For arbitrary positive integers n_0, m, $n_0<m$, we define the frames

$$F_j = \begin{cases} V_{n_0}, & j=0 \\ V_{2jR+n_0} \setminus V_{2(j-1)R+n_0}, & j=1, 2, \ldots, m \\ V^* = Z^2 \setminus V_{2jR+n_0}, & j = m+1 \end{cases}$$

We are going to verify the following inequality:

$$(3.3) \qquad |p(g\varphi(F_0)|\varphi(F_{m+1})) - p(\varphi(F_0)|\varphi(F_{m+1}))| \leqq \leqq Km^{-\gamma} p(\varphi(F_0)|\varphi(F_{m+1}))$$

for arbitrary $\varphi(F_0)$, $\varphi(F_{m+1})$ and $g \in G$. Here K and γ are appropriate positive constants independent of the configuration and m; $p(\cdot|\cdot)$ denotes the conditional Gibbs density function. The relation (3.3) implies the theorem. Indeed, by integrating (3.3), we obtain that, for any $g \in G$, $B \subset \Omega(V_{n_0})$ and limit Gibbs distribution $\mathcal{P}$,

$$|\mathcal{P}(gB) - \mathcal{P}(B)| \leqq Km^{-\gamma}.$$

(Given a set $V \subset Z^2$, $\Omega(V)$ denotes the direct power of the space Φ to the parameter set V.)

Letting m tend to infinity, we obtain the relation $\mathcal{P}(gB) = \mathcal{P}(B)$, i.e. the invariance of $\mathcal{P}$ under the action of G. Thus the problem reduces to the proof of (3.3).

First we give a heuristic argument for the proof of (3.3). The idea is the following: We consider the density function of the conditional distribution of the configuration $\varphi(F_0)$ under the condition $\varphi(F_{m+1})$, where $\varphi(F_{m+1})$ is a fixed configuration. We want to prove that its value at the point $g\varphi(F_0)$ is almost independent of g. Actually, we prove the following stronger result. We consider the space $\mathscr{F}$ consisting of the points $f=(\varphi(F_0), \varphi(F_1), \ldots, \varphi(F_m))$. We can endow $\mathscr{F}$ with a σ-algebra in a natural way. Then the conditional Gibbs distribution with respect to the condition $\varphi(F_{m+1})$ is a probability measure on it. We can define a measurable partition ξ of $\mathscr{F}$ (under some mild topological conditions on Φ) by the help of the following equivalence relation: $f=(\varphi(F_0), \ldots, \varphi(F_m))\sim f'=(\varphi'(F_0), \ldots, \varphi'(F_m))$ if $g_1, \ldots, g_m \in G$ can be found so that $\varphi'(F_i)=g_i\varphi(F_i)$, $i=0, 1, \ldots, m$. If C_ξ is an element of the partition ξ, then $g_0, \ldots, g_m$ can be considered as the coordinates in C_ξ. The crucial point is that the conditional density of the coordinates $g_0, \ldots, g_m$ under the conditions C_ξ, $\varphi(F_{m+1})$ in C_ξ, can be expressed in the form

$$p(g_0\varphi(F_0), g_1\varphi(F_1), \ldots, g_m\varphi(F_m)|C_\xi, \varphi(F_{m+1})) =$$
$$= \bar{\varrho}(g_m|\varphi(F_{m+1})) \prod_{i=0}^{m-1} \varrho_i(g_i g_{i+1}^{-1})$$

(the factors on the right-hand side depend on C_ξ too, though this is not explicitly indicated). This means that the relative rotations $g_i g_{i+1}^{-1}$, $i=0, 1, \ldots, m-1$ and g_m are independent random variables (with respect to this conditional distribution). The simplest way to verify the above statement is to write down the relation

$$\frac{p(g_0\varphi(F_0), \ldots, g_m\varphi(F_m)|\varphi(F_{m+1}))}{p(\varphi(F_0), \ldots, \varphi(F_m)|\varphi(F_{m+1}))} =$$
$$= \exp\left[G_0(g_0 g_1^{-1}) + G_1(g_1 g_2^{-1}) + \ldots + G_{m-1}(g_{m-1} g_m^{-1}) + \tilde{G}(g_m)\right].$$

It will be shown that, on every element C_ξ of the partition ξ, the conditional distribution of the coordinate g_0 is almost independent of $\varphi(F_{m+1})$ and this will imply (3.3). It follows from the foregoing that this conditional distribution can be written as the distribution of the product of independent random variables with values on the unit circle G. The required property follows from a limit theorem for the product of independent random variables taking their values on the unit sphere. In this limit theorem the individual terms must have sufficiently large variance, and this is satisfied in the two-dimensional models.

In order to avoid some measure-theoretical problems in connection with the measurable partitions, we shall prove (3.3) in a somewhat different way. Let $\mathfrak{J}$ be the Borel σ-algebra on $G=S^1$, $\bar{\mu}$ the normed Haar measure on it. We define the measurable space $(T, \mathscr{L}, \nu)$, $T=\Omega(\tilde{V}_{2mR+n_0})\times G^m$, $\mathscr{L}=\mathscr{F}_{V_{2mR+n_0}}\times\mathfrak{J}^m$, $\nu=\mu_{2mR+n_0}\times\bar{\mu}^m$. G^m is the m-fold direct product of the group G; $\mathfrak{J}^m$ and $\bar{\mu}^m$ are the direct products of the corresponding σ-algebras and measures.

Let us introduce the notation $M=2mR+n_0$. We define the probability measure $\mathscr{P}(\cdot|\varphi(F_{m+1}))$ on (M, B) with the following density function with respect to the measure ν:

$$p(\varphi(V_M), g_0, \dots, g_m) =$$

$$= \frac{d\mathscr{P}\{\varphi(V_M), g_0, g_1, \dots, g_m|\varphi(F_{m+1})\}}{d\nu} =$$

$$= \frac{\exp\{-\mathscr{U}(g_0\varphi(F_1), g_1\varphi(F_1), \dots, g_m\varphi(F_m)|\varphi(F_{m+1}))\}}{\int_{\Omega(V_M)}\exp\{-\mathscr{U}(\varphi(V_M)|\varphi(F_{m+1}))\}\,d\mu_{V_M}(\varphi(V_M))}.$$

(Given a finite set $F\subset Z^2$, we define

$$\mathscr{U}(\varphi(F)|\varphi(Z^2\setminus F)) = \sum_{W_R(s)\cap F\neq\varnothing}\mathscr{U}(\varphi(W_R(s))).)$$

In order to see that $\mathscr{P}\{\cdot|\varphi(F_{m+1})\}$ really is a probability measure, one has to observe that the denominator is equal to

$$\int_T \exp\{-\mathscr{U}(g_1\varphi(F_1), \dots, g_m\varphi(F_m)|\varphi(F_{m+1}))\}\,d\nu.$$

From now on we omit $\varphi(F_{m+1})$ from $\mathscr{P}(\cdot|\varphi(F_{m+1}))$ and $p(\cdot|\varphi(F_{m+1}))$, and we will write simply $\mathscr{P}$ and p.

(The measure $\mathscr{P}$ can be interpreted as follows: let us consider the conditional Gibbs distribution on the configurations in $\Omega(V_M)$ under the condition $\varphi(F_{m+1})$ in $\Omega(Z^2-V_M)$, and let us choose $g_i\in G$, $i=0, 1, \dots, m$, at random. We rotate the configuration in F_i with g_i^{-1}, $i=0, 1, \dots, m$. Then $p(\varphi(V_M), g_0, g_1, \dots, g_m)$ denotes the density function of the probability distribution of the following random variable: The elements $g_0, \dots, g_m$ are chosen, and $\varphi(V_M)$ is the configuration after the rotations. The relation $p(g_0|\varphi(F_0))\sim p(g_0'|\varphi(F_0))$ means that $g\varphi(F_0)$ and $g'\varphi(F_0)$ have almost the same probability density under the original conditional Gibbs distribution $\mathscr{P}(\cdot|\varphi(F_{m+1}))$. This is what we want to prove, therefore we will study the conditional density $p(g_0|\varphi(F_0))$.)

It is easy to see that, for arbitrary $g_0\in G$,

$$p(g_0|\varphi(F_0)) = p(g_0\varphi(F_0)|\varphi(F_{m+1}))\times$$

$$\times\frac{\int_{\Omega(V_M)}\exp\{-\mathscr{U}(\varphi(V_M)|\varphi(F_{m+1}))\}\,d\mu_{V_M}(\varphi(V_M))}{\int_{\Omega(V_M\setminus F_0)\times G}\exp\{-\mathscr{U}(g_0\varphi(F_0), \varphi(V_M\setminus F_0)|\varphi(F_{m+1}))\}\,d\mu_{V_M\setminus F_0}(\varphi(V_M-F_0))\,d\bar\mu(g)}.$$

Therefore, (3.3) is equivalent to

$$|p(g_0|\varphi(F_0))-p(e|\varphi(F_0))| \leqq K_1 m^{-\gamma} p(e|\varphi(F_0)),$$

where e is the unit element in G.

We prove that

$$|p(g_0|(F_0))-1| \geqq K_2 m^{-\gamma}.$$

Since

$$p(g_0|\varphi(F_0)) = \int_{\Omega(V_M\setminus F_0)} p(g_0|\varphi(V_M))\,p(\varphi(V_M)|\varphi(F_0))\times$$

$$\times d\mu_{V_M\setminus F_0}(\varphi(V_M\setminus F_0)),$$

it is enough to prove that

$$|p(g_0|\varphi(V_M))-1| \leqq K_2 m^{-\gamma}. \tag{3.4}$$

In order to prove the last relation, we consider the conditional density $p(g_0, \ldots$ $\ldots, g_m|\varphi(V_M))$. First, we compute

$$\mathscr{U}(g_0\varphi(F_0), g_1\varphi(F_1), \ldots, g_s\varphi(F_s)|\varphi(F_{s+1})).$$

Let us observe that, since the radius of the interaction is R, $\mathscr{U}(\varphi(V))\neq 0$ implies that either $V\subset F_i$ or $V\subset F_i\cup F_{i+1}$ for some $i=1, 2, \ldots, m$. If $V\subset F_i$, then $\mathscr{U}(\varphi(V))=$ $=\mathscr{U}(g\varphi(V))$.

$$\begin{aligned}\mathscr{U}(g_0\varphi(F_0), g_1\varphi(F_1), \ldots, g_m\varphi(F_m)|\varphi(F_{m+1})) &= \\ = \sum_{j=0}^m \sum_{s\in F_i} \mathscr{U}(\varphi(W_R(s))) + \sum_{\substack{s: W_R(s)\cap V_M\neq\varnothing \\ s\notin V_M}} \mathscr{U}(\varphi(W_R(s))) &= \\ = \sum_{j=0}^m U_j(\varphi(F_j)) + \sum_{j=1}^m R_j(g_j g_{j+1}^{-1}\varphi(F_j), \varphi(F_{j+1})),\end{aligned} \tag{3.5}$$

where $g_{m+1}=e$, and U_j and R_j, the summands in the first and in the second sum, only depend on the variables in the corresponding brackets.

Put $h_i=g_i g_{i+1}^{-1}$, $i= 0, 1, \ldots, m$ (setting $g_{m+1}=e$). We shall show that

$$\left|\frac{\partial^2}{\partial h_j^2} R_j(h_j\varphi(F_j), \varphi(F_{j+1}))\right| \leqq c_1 j \tag{3.6}$$

with some $c_1=c_1(d, L)$.

Indeed, every term in the sum expressing $R_j(h_j\varphi(F_j), \varphi(F_{j+1}))$ has the form $\mathscr{U}(\varphi(W_R(s)))$, where $W_R(s)\cap F_j\neq\varnothing$, $W_R(s)\cap F_{j+1}\neq\varnothing$. Differentiating each term twice, we get a sum with less than $8jR^4$ terms, and each term is bounded by L because of condition A).

Now we have the formula

$$\begin{aligned} p(g_0, \ldots, g_m|\varphi(V_M)) &= \\ = \frac{\exp\{-\mathscr{U}(g_0\varphi(F_0), \ldots, g_m\varphi(F_m)|\varphi(F_{m+1}))\}}{\int_{G^{m+1}}\exp\{-\mathscr{U}(g_0\varphi(F_0), \ldots, g_m\varphi(F_m)|\varphi(F_{m+1}))\}\,d\bar\mu(g_0)\ldots d\bar\mu(g_m)} &= \\ = \frac{\exp\{-\sum_{j=0}^m R_j(h_j\varphi(F_j), \varphi(F_{j+1}))\}}{\int_{G^{m+1}}\exp\{-\sum_{j=0}^m R_j(h_j\varphi(F_j), \varphi(F_{j+1}))\}\,d\bar\mu(h_0)\ldots d\bar\mu(h_m)} &= \\ = \prod_{j=0}^m \varrho_j(h_j|\varphi(F_j), \varphi(F_{j+1})), \end{aligned}$$

where

$$\varrho_j(h|\varphi(F_j), \varphi(F_{j+1})) = \frac{\exp\{-R_j(h\varphi(F_j), \varphi(F_{j+1}))\}}{\int_G \exp\{-R_j(h\varphi(F_j), \varphi(F_{j+1}))\}\,d\bar\mu(h)}.$$

We can see that under the condition $\varphi(V_M)$ the variables $h_i=g_i g_{i+1}^{-1}$, $i=0, 1, \ldots, m$, are conditionally independent with probability density functions $\varrho_i(h|\varphi(F_i), \varphi(F_{i+1}))$, and $g_0 = h_0 \dots h_m$.

Thus

$$p(g_0|\varphi(V_M)) =$$
$$= \varrho_0(\cdot|\varphi(F_0), \varphi(F_1)) * \varrho_1(\cdot|\varphi(F_1), \varphi(F_2)) * \dots * \varrho_m(\cdot|\varphi(F_m), \varphi(F_{m+1}))(g_0),$$

where $*$ denotes convolution on the unit circle. Let us now estimate $\max_{h \in G} \varrho_j(h|\varphi(F_j), \varphi(F_{j+1}))$. Let R_j take its minimum at v_j, i.e. let $R_j(v_j \varphi(F_j), \varphi(F_{j+1})) = \min_{h \in G} R_j(h\varphi(F_j), \varphi(F_{j+1}))$. Applying the Taylor formula in the Lagrange form and the estimate (3.6), we obtain that

$$R_j(h\varphi(F_j), \varphi(F_{j+1})) \leqq R_j(v_j \varphi(F_j), \varphi(F_{j+1})) + \frac{cj_1}{2}|h - v_j|^2,$$

$$\int_G \exp[-R_j(h\varphi(F_j), \varphi(F_{j+1}))]\,\bar{\mu}(dh) \geqq$$
$$\geqq \exp[-R_j(v_j \varphi(F_j), \varphi(F_{j+1}))] \cdot \int_0^1 \exp[-c_1 j|y - v_j|^2]\,dy \geqq$$
$$\geqq \exp[-R_j(v_j \varphi(F_j), \varphi(F_{j+1}))c_3^{-1} j^{-1/2}.$$

Therefore

$$\varrho_j(h|\varphi(F_j), \varphi(F_{j+1})) \leqq c_3 \sqrt{j}.$$

Now because of the last estimate and (3.7) the following lemma implies (3.4).

Lemma. *Let $\varrho_1(g), \dots, \varrho_m(g)$ be density functions on the unit circle G. Let us assume that $\varrho_j(g) \leqq c\sqrt{j}$, $j = 1, 2, \dots, m$ with some constant c. Then the density function $\varrho(g) = \varrho_1 * \dots * \varrho_m(g)$ satisfies the inequality $\sup_{g \in G} |\varrho(g) - 1| \leqq K_3 m^{-\gamma}$, where $\gamma = \frac{2}{3}\left(\frac{\pi}{c}\right)^2$, and K_3 also depends on c only.*

Proof of the Lemma. The proof is a natural adaptation of the characteristic function technique to our case. We may assume that $G = [0, 1)$ with addition modulo 1.

Let us consider the Fourier expansion

$$\varrho_j(x) = 1 + \sum_{k=1}^{\infty} (a_{k,j} e^{i2k\pi x} + a_{-k,j} e^{-i2k\pi x}), \quad j = 1, \dots, m.$$

Then the Fourier series of $\varrho(x)$ has the form

$$\varrho(x) = 1 + \sum_{k=1}^{\infty} (A_k e^{i2k\pi x} + A_{-k} e^{-i2k\pi x})$$

with $A_k = \prod_{j=1}^{m} a_{k,j}$.

We estimate the Fourier coefficients $a_{k,j}$ for $k \neq 0$.

$$|a_{k,j}| = \left|\int_0^1 e^{-i2k\pi x} \varrho_j(x)\,dx\right| = \left|\int_0^1 e^{-i2k\pi x} \varrho_j(x + \alpha_{j,k})\,dx\right| =$$
$$= \left|1 - \int_0^1 [1 - \cos 2k\pi x]\,\varrho_j(x + \alpha_{j,k})\,dx\right| \leqq 1 - c_j,$$

where $c_j = \frac{2\pi^2}{3c^2 j} - \frac{2\pi^4}{15c^4 j^2}$, and $\alpha_{j,k}$ is defined by the relation

$$\int_0^1 \sin(2k\pi x)\,\varrho_j(x+\alpha_{j,k})\,dx = 0.$$

The last inequality holds, because for the following class $\mathscr{S}$ of functions

$$\mathscr{S} = \left\{\varrho(x)\colon\ 0 \leqq \varrho(x) \leqq c\sqrt{j},\ x \in [0, 1],\ \int_0^1 \varrho(x)\,dx = 1\right\},$$

we have

$$\min_{\varrho\in\mathscr{S}} \int_0^1 [1-\cos 2k\pi x]\,\varrho(x)\,dx = \int_0^1 [1-\cos 2k\pi x]\,\varrho^*(x)\,dx,$$

where

$$\varrho^*(x) = \begin{cases} c\sqrt{j} & \text{if } x \in \left[\frac{i}{k}, \frac{i}{k} + \frac{1}{ck\sqrt{j}}\right], \quad i = 0, 1, \dots, k-1 \\ 0 & \text{otherwise.} \end{cases}$$

By the Parseval relation, we have

$$1 + \sum_{k=1}^{\infty} (|a_{k,j}|^2 + |a_{-k,j}|^2) = \int_0^1 \varrho_j^2(x)\,dx \leqq c\sqrt{j}.$$

Put

$$A_{k,l} = \prod_{j=1}^{l} a_{k,j}, \qquad -\infty < k < \infty,\ k \neq 0,\ l = 1, 2, \dots, m$$

and

$$A_{0,l} = 0.$$

We have

$$\sum_{k=-\infty}^{\infty} |A_{k,2}| \leqq \left(\sum |a_{k,1}|^2\right)^{1/2} \left(\sum |a_{k,2}|^2\right)^{1/2} \leqq 2c.$$

If $l > 2$, we obtain by induction

$$\sum_{k=-\infty}^{\infty} |A_{k,l}| \leqq (1-c_l) \sum_{k=-\infty}^{\infty} |A_{k,l-1}| \leqq 2c \prod_{j=3}^{\infty} (1-c_j) \leqq$$

$$\leqq K_4 \exp\left(-\sum_{j=1}^{l} \frac{2\pi^2}{3c^2 j}\right) \leqq K_5\, l^{-2\pi^2/3c^2}.$$

Since $\sum |A_k| < \infty$, $\varrho(x)$ is a continuous function, and $\varrho(x) - 1 = \sum A_k e^{i2k\pi x}$ for every x. Thus

$$|\varrho(x) - 1| \leqq \sum_{k=1}^{\infty} (|A_k| + |A_{-k}|) \leqq \sum_{k=-\infty}^{\infty} |A_{k,m}| \leqq K_3\, m^{-2\pi^2/3c^2}$$

as we claimed. Thus the proof of relation (3.3) and also the proof of the theorem is completed. Q.e.d.

By the help of the above theorem the absence of breakdown of continuous symmetry can be proved for some two-dimensional models with other symmetry groups.

Let G be a torus, i.e. the direct product of finitely many circles, $G = G_1 \times G_2 \times \dots \times G_r$. Let G act on the space Φ. Since the Hamiltonian is invariant under each G_i, it follows from the above theorem that every limit Gibbs distribution $\mathscr{P}_\beta$ is invariant under G.

Using this remark we show that in the two-dimensional Heisenberg model no spontaneous breakdown of symmetry occurs, i.e. there is no limit Gibbs distribution non-invariant under the action of the group G.

In this model $\Phi = S^{\nu-1}$, the unit sphere in the ν-dimensional space, μ is the normed Lebesgue measure on $S^{\nu-1}$, and the potential is given by the formula

$$\mathscr{U}(\varphi(W(s))) = \begin{cases} 1 - \frac{1}{4}(\varphi(s), \varphi(s')) & \text{if} \quad |s-s'| = 1 \\ 0 \quad \text{otherwise.} \end{cases}$$

The measure μ and the potential $\mathscr{U}$ are invariant under the actions of the group $G = SO(\nu)$, the group of the rotations in the ν-dimensional euclidean space R^ν. We prove the following consequence of the theorem.

Corollary. *In the two-dimensional Heisenberg model every limit Gibbs distribution is invariant under the group $G = SO(\nu)$ at arbitrary $\beta > 0$.*

Proof. We want to show that $\mathscr{P}_\beta$ is invariant under the action of an arbitrary $A \in G$. It is well known from linear algebra that, in an appropriate coordinate system, A can be decomposed into the direct product of 2×2 and 1×1 matrices with determinant 1. This is equivalent to saying that R^ν can be decomposed into the direct sum of orthogonal subspaces, invariant under A, i.e. $R^\nu = B_1 \times \ldots \times B_s$, $\dim B_i \leqq 2$, and B_i is an A-invariant subspace of R^ν, and the restriction of A to B_i is a rotation. Let G_i be the group of rotations on B_i. Then $G^* = G_1 \times G_2 \times \ldots \times G_s$ is a torus containing A. The Hamiltonian of the Heisenberg model is invariant under each G_i, so $\mathscr{P}_\beta$ is invariant under A. Q.e.d.

It is well known that, if G is a compact connected Lie group, and $g \in G$, then there exists a Cartan subgroup $G_0 \subset G$ with $g \in G_0$, isomorphic to a torus. Thus the theorem can also be generalized to compact connected Lie groups. The condition about the finite interaction radius can be weakened to exponentially decreasing interaction. On the other hand, if the interaction decreases as a power of the distance, a breakdown of continuous symmetry may occur. The differentiability conditions are likely to be essential.

3. The Fröhlich–Simon–Spencer theorem on the existence of spontaneous magnetization in the d-dimensional classical Heisenberg model, $d \geqq 3$

We consider the d-dimensional classical Heisenberg model, $d \geqq 3$. The spin variables $\varphi(s)$ take values in $S^{\nu-1}$ and the Hamiltonian is

$$\mathscr{H} = -\sum_{\|s_1 - s_2\| = 1} (\varphi(s_1), \varphi(s_2)).$$

Let $V = V_n$ denote the d-dimensional discrete torus with edges of length n ($|V| = n^d$), i.e. let V_n be the d-direct power of the group $G=\{1, 2, \ldots, n\}$, where the group operation is the addition modulo n. V can be embedded into the lattice Z^d in a natural way. We shall study the probability distribution on the configurations $\varphi(V)$ generated by the Hamiltonian $\mathscr{H}$ with periodic boundary condition at an inverse temperature β.

This means the following: μ denotes the Lebesgue measure on $S^{\nu-1}$. Then the Gibbs probability distribution with periodic boundary conditions is given by the formula

$$dP_V = \frac{e^{-\beta\mathscr{H}_V}}{\Theta_V} \prod_{s\in V} d\mu_s,$$

where $\mathscr{H}_V = -\sum_{\substack{\|s_2-s_1\|=1 \\ s_1\in V, s_2\in V}} (\varphi(s_1), \varphi(s_2))$, μ_s is the Lebesgue measure on the sphere where the variables $\varphi(s)$ take its values.

Put $\varphi_V = \frac{1}{|V|}\sum_{s\in V} \varphi(s)$. Then φ_V is a d-dimensional vector.

Theorem 3.2. (Fröhlich, Simon and Spencer [64–67]). *There exists a* $\beta_0=\beta_0(\nu, d)>0$ *such that*

$$M = \liminf_{V\to\infty} E(\varphi_V, \varphi_V) > 0$$

whenever $\beta>\beta_0$. *Here E stands for mathematical expectation with respect to the measure* P_V.

Remarks. 1) It follows from the proof that β_0 can be chosen as

$$\beta_0 = \frac{\nu}{2(2\pi)^d}\int_{-\pi}^{\pi}\cdots\int_{-\pi}^{\pi}\left[\sum_{k=1}^{d}(1-\cos p_k)\right]^{-1} dp_1\ldots dp_d.$$

2) The theorem can be generalized to systems with Hamiltonian of the form

$$\mathscr{H}_v = -\sum_{s_1, s_2\in V} U(s_1-s_2)(\varphi(s_1), \varphi(s_2)),$$

where U is a potential with finite radius of interaction such that $\sum_{t\neq 0} U(t)>0$, $U(t)\geqq 0$, $U(t)=U(t_1, \ldots, t_d)=U(\pm t_1, \ldots, \pm t_d)$

Proof*. Let $V=V_n$, $p=\frac{2\pi}{n}q$, $q\in V$,

$$\tilde{\varphi}^{(m)}(q) = |V|^{-1/2}\sum_{s\in V}\exp\{i(s, p)\}\cdot\varphi^{(m)}(s),$$

where $\varphi^{(m)}(s)$ is the mth coordinate of $\varphi(s)\in S^{\nu-1}\subset R^{\nu}$, $1\leqq m\leqq \nu$.

We need the following

* The present proof was explained by B. Simon at the IVth International Symposium on Information Theory held in June 1976 in Repino. The details of the proof were worked out by P. M. Bleher. Our text closely follows his proof.

Basic Lemma.

$$E|\tilde{\varphi}^{(m)}(p)|^2 \leqq \frac{1}{2}(\beta E_p)^{-1},$$

where $E_p=\sum_{j=1}^{d}(1-\cos p_j)$, $m=1, 2, \ldots, \nu$.

Proof of the Theorem by means of the Basic Lemma. We have

$$|\varphi_V|^2 \cdot |V| = |\tilde{\varphi}_0|^2 = \sum_{m=1}^{\nu} |\tilde{\varphi}^{(m)}(0)|^2.$$

It follows from the Parseval equality that

$$\sum_{p\in V^*} \|\tilde{\varphi}(p)\|^2 = \sum_{s\in V} \|\varphi(s)\|^2,$$

where $V^* = \left\{p : p = \frac{2\pi}{n} q,\ q\in V\right\}$.

The right-hand side of this equation equals $|V|$ since $\|\varphi(s)\|=1$. As a result

$$|V| = \sum_{\substack{p\in V^*\\ p\neq 0}} \|\tilde{\varphi}(p)\|^2 + \|\tilde{\varphi}(0)\|^2 = \sum_{\substack{p\in V^*\\ p\neq 0}} \|\tilde{\varphi}(p)\|^2 + |V|\cdot|\varphi_V|^2.$$

Taking the expectation with respect to the measure P_V on both sides of the equation and applying the Basic Lemma, we obtain that

$$|V|(1-E|\varphi_V|^2) < \frac{\nu}{2} \sum_{\substack{p\in V^*\\ p\neq 0}} (\beta E_p)^{-1}.$$

For $d \geqq 3$ the integral $\int_{-\pi}^{\pi} \cdots \int_{-\pi}^{\pi} E_p^{-1} dp_1 \ldots dp_d$ is convergent, and

$$|V|^{-1} \sum_{\substack{p\in V^*\\ p\neq 0}} E_p^{-1} \to \frac{1}{(2\pi)^d} \int_{-\pi}^{\pi} \cdots \int_{-\pi}^{\pi} E_p^{-1}\, dp_1 \ldots dp_d.$$

Thus

$$\limsup_{V\to\infty} (1-E|\varphi_V|^2) \leqq c_0\beta^{-1} \quad \text{with} \quad c_0 = \frac{\nu}{2(2\pi)^d} \int_{-\pi}^{\pi} \cdots \int_{-\pi}^{\pi} E_p^{-1}\, dp.$$

In other words

$$\liminf_{V\to\infty} |\varphi_V|^2 > 1 - c_0\beta^{-1}.$$

Choosing $\beta_0 = c_0$ we get the statement of the Theorem.

Proof of the Basic Lemma. Let V_Γ denote the set of edges connecting the neighbouring points of V. Each edge can be represented as a pair (s, k), where $s\in V$, and k corresponds to the unit vector $\delta_k = \{\underbrace{0, \ldots, 0}_{k-1}, 1, 0, \ldots, 0\}$. Let $h(s, k)$ be an arbitrary function on V_Γ with values in R^1 and let λ be a real number. For every m, $1 \leqq m \leqq \nu$, we define the function

$$I_\lambda(\{h(s, k)\}) = E\left[\exp\left(\lambda \sum_{(s,k)\in V_\Gamma} h(s, k)(\varphi^{(m)}(s+\delta_k) - \varphi^{(m)}(s))\right)\right].$$

The following *basic estimate* holds:

$$I_\lambda(\{h(s,k)\}) \leqq \exp\left\{\frac{1}{2\beta}\lambda^2 \sum_{(s,k)\in V_\Gamma} |h(s,k)|^2\right\}.$$

Derivation of the Basic Lemma from the basic estimate: We have

$$\frac{d}{d\lambda} I_\lambda(\{h(s,k)\})\big|_{\lambda=0} = \sum h(s,k)\, E(\varphi^{(m)}(s+\delta_k)-\varphi^{(m)}(s)) = 0$$

because of the symmetric boundary conditions. Therefore it follows from the basic estimate that

$$\frac{d^2}{d\lambda^2} I_\lambda(\{h(s,k)\})\big|_{\lambda=0} \leqq \frac{d^2}{d\lambda^2}\exp\{(2\beta)^{-1}\lambda^2 \sum_{(s,k)\in V_\Gamma} |h(s,k)|^2\}\big|_{\lambda=0}$$

or in an other form

(3.8)
$$E\left|\sum_{(s,k)\in V_\Gamma} h(s,k)\big(\varphi^{(m)}(s+\delta_k)-\varphi^{(m)}(s)\big)\right|^2 \leqq$$
$$\leqq \beta^{-1}\sum_{(s,k)\in V_\Gamma} |h(s,k)|^2.$$

We apply the last inequality by choosing $h(s,k)=\operatorname{Re}(e^{i(p,s+\delta_k)}-e^{i(p,s)})$. We may write

$$\begin{aligned}
&\sum_{(s,k)\in V_\Gamma} h(s,k)\big(\varphi^{(m)}(s+\delta_k)-\varphi^{(m)}(s)\big) = \\
&= \sum_{k=1}^{d}\sum_{s\in V}[h(s-\delta_k,k)-h(s,k)]\,\varphi^{(m)}(s) = \\
&= \operatorname{Re}\sum_{k=1}^{d}\sum_{s\in V}(2e^{i(p,s)}-e^{i(p,s+\delta_k)}-e^{i(p,s-\delta_k)})\,\varphi^{(m)}(s) = \\
&= \sum_{k=1}^{d}(2-e^{ip_k}-e^{-ip_k})\cdot\operatorname{Re}\sum_{s\in V} e^{i(p,s)}\varphi^{(m)}(s) = \\
&= 2|V|^{1/2}\sum_{k=1}^{d}(1-\cos p_k)\operatorname{Re}\tilde\varphi^{(m)}(p) = 2E_p|V|^{1/2}\operatorname{Re}\tilde\varphi^{(m)}(p).
\end{aligned}$$

Thus, the left-hand side of (3.8) equals

$$L = 4|V|\cdot E_p^2 E(\operatorname{Re}\tilde\varphi^{(m)}(p))^2,$$

and we get that

$$4|V|\,E_p^2 E(\operatorname{Re}\tilde\varphi^{(m)}(p))^2 \leqq \beta^{-1}\sum_{s,k}\operatorname{Re}(e^{i(p,s+\delta_k)}-e^{i(p,s)})^2.$$

An analogous consideration shows that, choosing $h(s,k)=\operatorname{Im}(e^{i(p,s+\delta_k)}-e^{i(p,s)})$, in the inequality (3.8), we get the estimate

$$4|V|E_p^2 E(\operatorname{Im}\tilde\varphi^{(m)}(p))^2 \leqq \beta^{-1}\sum_{s,k}\operatorname{Im}(e^{i(p,s+\delta_k)}-e^{i(p,s)})^2.$$

By combining these estimates we obtain

$$4|V|\,E_p^2 E|\tilde\varphi^{(m)}(p)|^2 \leqq \beta^{-1}\sum_{(s,k)\in V_\Gamma}|e^{i(p,s+\delta_k)}-e^{i(p,s)}|^2 =$$
$$= \beta^{-1}\sum_{(s,k)\in V_\Gamma}|e^{ip_k}-1|^2 = 2\beta^{-1}|V|\sum_{k=1}^{d}(1-\cos p_k) = 2\beta^{-1}|V|E_p.$$

Thus

$$E|\tilde{\varphi}^{(m)}(p)|^2 \leqq \frac{1}{2}\beta^{-1}E_p^{-1},$$

and the Basic Lemma is proven.

Proof of the Basic Estimate.

Writing $h(s, k)$ instead of $\lambda h(s, k)$ we may reduce the problem to the special case $\lambda=1$. Furthermore, we have by definition

$$I_1(\{h(s, k)\}) = Z'^{-1}\int \exp\{\beta \textstyle\sum_{s,k}(\varphi(s), \varphi(s+\delta_k))+$$

$$+\textstyle\sum_{s,k} h(s, k)[\varphi^{(m)}(s+\delta_k)-\varphi^{(m)}(s)]\}\prod d\mu(\varphi(s))$$

with

$$Z' = \int \exp\{\beta \textstyle\sum_{s,k}(\varphi(s), \varphi(s+\delta_k))\}\prod d\mu(\varphi(s)).$$

Since

$$-(\varphi(s), \varphi(s+\delta_k)) = \frac{1}{2}(\varphi(s+\delta_k)-\varphi(s), \varphi(s+\delta_k)-\varphi(s))-$$

$$-\frac{1}{2}(\varphi(s+\delta_k), \varphi(s+\delta_k))-\frac{1}{2}(\varphi(s), \varphi(s)) = \frac{1}{2}\|\varphi(s+\delta_k)-\varphi(s)\|^2-1,$$

the quantity $I_1(\{h(s, k)\})$ can be written as

$$I_1(\{h(s, k)\}) = Z^{-1}\int \exp\{-\frac{\beta}{2}\textstyle\sum_{s,k}\|\varphi(s+\delta_k)-\varphi(s)\|^2+$$

$$+\textstyle\sum_{s,k}(\bar{h}(s, k), \varphi(s+\delta_k)-\varphi(s))\}\prod d\mu(\varphi(s)),$$

where

$$Z = \int \exp\left\{-\frac{\beta}{2}\textstyle\sum_{s,k}\|\varphi(s+\delta_k)-\varphi(s)\|^2\right\}\prod d\mu(\varphi(s)),$$

and $\bar{h}(s, k)$ denotes the vector $(\underbrace{0, \dots, 0}_{m-1}, h(s, k), 0, \dots, 0)$. Thus

$$I_1(\{h(s, k)\}) = Z^{-1}\int \exp\left\{-\frac{\beta}{2}\textstyle\sum_{s,k}\|\varphi(s+\delta_k)-\varphi(s)-\frac{1}{\beta}\bar{h}(s, k)\|^2\right\}\prod d\mu(\varphi(s))\times$$

$$\times\mathscr{I}(\{h(s, k)\})$$

with

$$\mathscr{I}(\{h(s, k)\}) = \exp\left(\frac{1}{2\beta}\textstyle\sum_{s,k}|h(s, k)|^2\right).$$

The expression $\mathscr{I}(\{h(s, k)\})$ coincides with the right-hand side of the basic estimate.

Thus, the basic estimate will be proved if we show that

$$Z^{-1}\int \exp\left\{-\frac{\beta}{2}\sum_{s,k}\|\varphi(s+\delta_k)-\varphi(s)-\frac{1}{\beta}\bar{h}(s,k)\|^2\right\}\Pi\, d\mu(\varphi(s)) \leqq 1.$$

Introducing the notation

$$Z(\{h(s,k)\}) = \int \exp\left\{-\frac{\beta}{2}\sum\|\varphi(s+\delta_k)-\varphi(s)-\frac{1}{\beta}\bar{h}(s,k)\|^2\right\}\Pi\, d\mu(\varphi(s))$$

the last inequality can be written in the form

$$Z(\{h(s,k)\}) \leqq Z(\{0\}), \tag{3.9}$$

i.e. we have to prove that the maximum of the statistical sum $Z(\{h(s,k)\})$ is reached at $h\equiv 0$.

Let us introduce the notation

$$V^{(r)} = \{s=(s_1\dots s_d)\in V\,|\,s_1=r\},\quad r=1,\dots,n,$$

$$\varphi = \{\varphi(s), s\in V\},\quad \varphi^{(r)} = \{\varphi(s), s\in V^r\}.$$

The set of possible configurations for a fixed r is a manifold $M=\mathop{\times}_{s\in V^{(r)}} S^{\nu-1}$, where the measure $d\mu(\varphi)=\Pi_{s\in V^{(r)}}d\mu_s$ can be introduced. In the space $\mathscr{L}^2(M,\mu(d\varphi))$, let us consider the operators $\hat{F}_1, \hat{T}_1, \dots, \hat{F}_n, \hat{T}_n$ defined in the following way:

The operator $\hat{F}_r$ acts as multiplication given by the function

$$F_r(\varphi^{(r)}) = \exp\left\{-\frac{\beta}{2}\sum_{\substack{s\in V^r\\ k\neq 1}}\|\varphi(s+\delta_k)-\varphi(s)-\beta^{-1}h(s,k)\|^2\right\}.$$

The operator $\hat{T}_r$ is an integral operator given by the kernel

$$T_r(\varphi^{(r)},\varphi^{(r+1)}) = \exp\left(-\frac{\beta}{2}\sum_{s\in V^r}\|\varphi(s+\delta_1)-\varphi(s)-\beta^{-1}h(s,1)\|^2\right).$$

It is not difficult to see that

$$\begin{aligned} Z(\{h(s,k)\}) &= \\ &= \int\dots\int d\mu(\varphi^{(1)}),\dots,d\mu(\varphi^{(n)})\,F(\varphi^{(1)})\,T_1(\varphi^{(1)},\varphi^{(2)})\dots \\ &\dots F_n(\varphi^{(n)})\,T_n(\varphi^{(n)},\varphi^{(1)}) = \mathrm{Tr}(\hat{F}_1\hat{T}_1\dots\hat{F}_n\hat{T}_n). \end{aligned} \tag{3.10}$$

Let the integral operator $\hat{T}_0$ be given by the kernel

$$T_0(\varphi^{(1)},\varphi^{(2)}) = \exp\left(-\frac{\beta}{2}\sum_{s\in V^{(1)}}|\varphi(s+\delta_1)-\varphi(s)|^2\right),$$

where $\{\varphi(s)\}=\varphi^{(1)}$ and $\{\varphi(s+\delta_1)\}=\varphi^{(2)}$.

We shall see in Lemma 3.1 below that

$$\hat{T}_0 > 0.$$

We may write

$$Z(\{h(s,k)\}) = \mathrm{Tr}\,(\hat{F}_1\hat{T}_1 \dots \hat{F}_n\hat{T}_n) = \tag{3.11}$$

$$= \mathrm{Tr}\,(\hat{T}_0^{1/2}\hat{F}_1\hat{T}_0^{1/2}\hat{T}_0^{-1/2}\hat{T}_1\hat{T}_0^{-1/2}\hat{T}_0^{1/2}\hat{F}_2\hat{T}_0^{1/2}\dots\hat{T}_0^{1/2}\hat{F}_n\hat{T}_0^{1/2}\hat{T}_0^{-1/2}\hat{T}_n\hat{T}_0^{-1/2}) =$$

$$= \mathrm{Tr}\,(A_1B_1 \dots A_nB_n),$$

where

$$A_r = \hat{T}_0^{1/2}\hat{F}_r\hat{T}_0^{1/2}, \quad B_r = \hat{T}_0^{-1/2}\hat{T}_r\hat{T}_0^{-1/2}.$$

Now we formulate three lemmas.

Lemma 3.1.

$$\hat{T}_0 > 0, \quad A_r > 0, \quad r = 1, 2, \dots, n.$$

Lemma 3.2.

$$\mathrm{Tr}\;(A_1B_1\dots A_nB_n) \leqq \|B_1\| \dots \|B_n\| \cdot (\mathrm{Tr}\,A_1^n)^{1/n} \dots (\mathrm{Tr}\,A_n^n)^{1/n}.$$

Lemma 3.3.

$$\|B_r\| \leqq 1, \quad r = 1, \dots, n.$$

We postpone the proof of these lemmas for a while and now we complete the proof of the basic estimate. It follows from Lemmas 3.2 and 3.3 that

$$Z(\{h(s,k)\}) \leqq (\mathrm{Tr}\,A_1^n)^{1/n} \dots (\mathrm{Tr}\,A_n^n)^{1/n},$$

thus the required inequality (3.9) will be proved if we show that

$$\mathrm{Tr}\,A_r^n \leqq Z(\{0\}), \quad r = 1, \dots, n.$$

Let us write

$$\mathrm{Tr}\,A_r^n = \mathrm{Tr}\,(\hat{T}_0^{1/2}\hat{F}_r\hat{T}_0^{1/2}\dots\hat{T}_0^{1/2}\hat{F}_r\hat{T}_0^{1/2}) =$$

$$= \mathrm{Tr}\,(\hat{F}_r\hat{T}_0\hat{F}_r\hat{T}_0\dots\hat{F}_r\hat{T}_0).$$

Comparing this equation with (3.10) we can see that

$$\mathrm{Tr}\,A_r^n = Z(\{\tilde{h}(s,k)\}),$$

where $\tilde{h}(s,1)=0$ for every $s \in V$, and $\tilde{h}(s,k)=h(s',k)$ with $s'=(r, s_2, \dots, s_d)$ if $k \neq 1$.

Thus our problem is to prove the inequality

$$Z(\{\tilde{h}(s,k)\}) \leqq Z(\{0\}), \tag{3.12}$$

which is analogous to the inequality (3.9). In this way the proof of (3.9) for an arbitrary set $\{h(s,k)\}$ is reduced to its proof for the special sets $\{\tilde{h}(s,k)\}$ having the property $\tilde{h}(s,1)=0$ for every $s \in V$. Repeating this argument we get to an analogous inequality for the sets $\{\tilde{\tilde{h}}(s,k)\}$ satisfying $\tilde{\tilde{h}}(s,1)=\tilde{\tilde{h}}(s,2)=0$ for every $s \in V$, etc. It is clear that choosing all the possible directions of the axes successively, we finally obtain the inequality (3.9). Thus the basic lemma follows from lemmas 3.1, 3.2 and 3.3.

Proof of Lemma 3.1. As

$$T_0(\varphi^{(1)}, \varphi^{(2)}) = \exp\left\{-\frac{\beta}{2} \sum_{s\in V^{(1)}} |\varphi^{(1)}(s)-\varphi^{(2)}(s)|^2\right\} =$$

$$= T_0(\varphi^{(2)}, \varphi^{(1)}),$$

the operator $\hat{T}_0$ is symmetrical.

We show that

$$(\hat{T}_0 f, f) \geqq 0$$

for any $f \in \mathscr{L}^2(M, \mu(d\varphi))$, and it is positive if $f \neq 0$ almost everywhere with respect to the measure μ.

The manifold M can be embedded into the space $M' = \prod_{s\in V^{(r)}} R^d$.

$C \exp\left\{-\frac{\beta}{2} |\varphi^{(1)}|^2\right\}$ is a normal density function on M' if $C > 0$ is appropriately chosen. Thus it can be expressed as the inverse Fourier transform of its characteristic function. This formula yields that

$$T_0(\varphi^{(1)}, \varphi^{(2)}) = c' \int_{M'} \exp[i(u, \varphi^{(1)} - \varphi^{(2)})] \exp\left[-\frac{1}{2\beta} |u|^2\right] du$$

with some $c' > 0$.

By means of this formula $(\hat{T}_0 f, f)$ can be written as

$$\int_M \int_M \int_{M'} \exp\left[-i(u, \varphi^{(2)} - \varphi^{(1)}) - \frac{1}{2\beta} |u|^2\right] du f(\varphi^{(1)}) f(\varphi^{(2)})\, d\mu(\varphi^{(1)})\, d\mu(\varphi^{(2)}) =$$

$$= \int_{M'} \left|\int_M \exp[-i(u, \varphi)] f(\varphi)\, d\mu(\varphi)\right|^2 \exp\left(-\frac{1}{2\beta} |u|^2\right) du.$$

The last expression is always non-negative, and it is strictly positive if $f(\varphi)$ is nonzero μ-almost everywhere. This means that $\hat{T}_0 > 0$. The second part of Lemma 3.1 is obvious since $\hat{F}_r$ is a multiplication operator with a positive function.

Proof of Lemma 3.2. We omit this proof since it can be found in the book of I. C. Gohberg and M. S. Krein *Introduction to the theory of linear non-self-adjoint operators* (in Russian; Nauka, Moscow 1965; formula (7.5) on p. 121 of this book).

Proof of Lemma 3.3. We have to show that

$$(\hat{T}_0^{-1/2} \hat{T}_r \hat{T}_0^{-1/2} f, f) \leqq f$$

or in an other form

$$(\hat{T}_r F, F) \leqq (\hat{T}_0 F, F) \tag{3.13}$$

where $F = \hat{T}_0^{-1/2} f$.

We may argue similarly as in the proof of Lemma 3.1. By a Fourier transform we can write (3.13) as

$$\int_{M'} \left| \int_M \exp[-i(u, \varphi)] F(\varphi) \, d\mu(\varphi) \right|^2 \exp\left(-\frac{1}{2\beta}|u|^2 - i\frac{1}{\beta}(h^r(s, 1), u)\right) du \leqq$$

$$\leqq \int_{M'} \left| \int_M \exp[-i(u, \varphi)] F(\varphi) \, d\mu(\varphi) \right|^2 \exp\left(-\frac{1}{2\beta}|u|^2\right) du$$

where $h^r(s, 1)$ is the restriction of $\{h(s, 1)\}$ to the set $s \in V^r$.

This inequality obviously holds since the outer integrand on the right-hand side is the absolute value of that on the left-hand side. Thus Lemma 3.3 is proved. Q.e.d.

Let us remark that there is a little inaccuracy in the above proof, but it can easily be improved. Namely, we cannot exclude the possibility that $\hat{T}_0^{-1/2}$ is an unbounded operator, and therefore not everywhere defined. This fact may cause some problems in the proof of Lemma 3.3. But one can get rid of these difficulties by approximating the Lebesgue measure μ on $S^{\nu-1}$ with measures μ_N concentrated on a finite subset of $S^{\nu-1}$. Redefining the operators $\hat{F}_r$, $\hat{T}_r$ and $\hat{T}_0$ by changing μ for μ_N we get a new version of Lemmas 3.1–3.3 which can be proved in the same way. The above-mentioned problem about the unboundedness of $\hat{T}_0^{-1/2}$ disappears since the new operators act on a finite-dimensional space. Finally, letting μ_N weakly tend to μ we can obtain the proof of the basic estimate.

Let us fix a positive integer N, and consider the rectangular N_d consisting of those points of V whose coordinates are between 0 and N. We define

$$\varphi_N = \frac{1}{N^d} \sum_{s \in V_d} \varphi(s).$$

We need the following simple

Lemma 3.4. For every $N \leqq n$ we have

$$E(\varphi_N, \varphi_N) \geqq E(\varphi_V, \varphi_V).$$

Proof. The identity $E(\varphi(s), \varphi(s+t)) = E(\varphi(0), \varphi(t))$ holds for every $s, t \in V$. Therefore $h(t) = E(\varphi(0), \varphi(t))$, $t \in V$, is a positive-definite function on the torus V. Thus we can write

$$E(\varphi(s), \varphi(t)) = \sum_{p \in V^*} e^{i(p, t-s)} \varrho(p),$$

where $\varrho(\cdot)$ is the spectral density of the positive-definite function $h(t)$. ($\varrho(p)$ can be given explicitly as $\varrho(p) = \frac{1}{|V|} E|\tilde{\varphi}(p)|^2$, but we do not need this fact.) By means of the last formula, we obtain that

$$E(\varphi_N, \varphi_N) = \frac{1}{N^{2d}} \sum_{p \in V^*} \left| \sum_{t \in N_d} e^{i(p, t)} \right|^2 \varrho(p) \geqq \varrho(0).$$

On the other hand

$$E(\varphi_V, \varphi_V) = \varrho(0).$$

These relations imply Lemma 3.4.

Remark. The correlation function $E(\varphi(t), \varphi(s))$ can be estimated in the following way:

$$E(\varphi(s), \varphi(t)) \geqq 1 - \frac{2c_0}{\beta} \quad \text{for every} \quad s, t \in V.$$

Proof.

$\varrho(0) = \frac{1}{V}(\varphi(0), \varphi(0)) \geqq 1 - \frac{c_0}{\beta}$. On the other hand $\sum_{p \in V^*} \varrho(p) = 1$. Therefore

$$E(\varphi(s), \varphi(t)) = \sum_{p \in V^*} e^{i(p, s-t)} \varrho(p) \geqq \varrho(0) - \sum_{\substack{p \neq 0 \\ p \in V^*}} \varrho(p) \geqq 1 - \frac{2c_0}{\beta}.$$

Now we briefly explain why the above theorem implies the existence of a limit Gibbs distribution with long-range order in the d-dimensional classical Heisenberg model, $d \geqq 3$.

Let $\mathscr{P}$ be a limit Gibbs distribution arising as a limit point of the measures $\mathscr{P}_V$ constructed before. (More precisely we can identify V_n with the subset $V_n = \prod_{i=1}^{d} \left\{ -\left[\frac{n}{2}\right], \dots, -\left[\frac{n}{2}\right] + n \right\}$ of Z^d, and consider an arbitrary measure $\tilde{P}_V$ on $\prod_{q \in Z^d} S^{\nu-1}(q)$ whose restriction to $\prod_{q \in \tilde{V}_n} S^{\nu-1}(q)$ agrees with P_V. We are speaking of limit point of these measures $\tilde{P}_V$. The measure $\mathscr{P}$ is invariant under the action of a multi-parameter dynamical system of space translations on the lattice Z^d and is also invariant under the action of the group $SO(\nu)$. It follows from the Fröhlich–Simon–Spencer theorem and Lemma 3.4 that, for $\beta > \beta_0$, $\lim_{V \to \infty} E \left| \frac{1}{V} \sum_{s \in V} \varphi(s) \right|^2 > 0$, where the expectation is taken with respect to the measure $\mathscr{P}$. By the multidimensional generalization of the Birkhoff–Khintchin ergodic theorem, the limit $\lim_{V \to \infty} \frac{1}{V} \sum_{s \in V} \varphi(s)$ exists with probability 1 with respect to the measure $\mathscr{P}$, and it follows from the previous remark that it is not equal to a constant vector with probability 1. Therefore $\mathscr{P}$ cannot be ergodic.

It is known from the general theory that every translation-invariant limit Gibbs state with a compact phase space is the mixture of extremal translation-invariant limit Gibbs states which are ergodic. Using this fact it can be shown that $\mathscr{P}$ is the mixture of the limit Gibbs states $\mathscr{P}_\varphi$, $\varphi \in R^\nu$, where $\mathscr{P}_\varphi$ is the conditional distribution of $\mathscr{P}$ under the condition that $\lim_{V \to \infty} \frac{1}{V} \sum_{s \in V} \varphi(s) = \varphi$. If $A \in SO(\nu)$, the rotation A

takes the Gibbs state $\mathscr{P}_\varphi$ to the state $\mathscr{P}_{A\varphi}$. If β is very large, then $\lim_{V\to\infty} E\left|\frac{1}{V}\sum_{s\in V}\varphi(s)\right|^2$ is almost 1. Thus there exist limit Gibbs states $\mathscr{P}_\varphi$ such that $|\varphi|$ is close to 1. Such a limit state can be considered as a small perturbation of the ground state $\varphi(s)\equiv\frac{\varphi}{|\varphi|}$, $s\in Z^d$.

Historical notes and references to Chapter III

1. The general question about the breakdown of continuous symmetry and the appearance of the so-called Goldstone particles is widely discussed in the physics literature. We mention the lectures of Berestecky [4] among the many examples.

2. The absence of long-range order in some two-dimensional classical models with continuous symmetry was obtained by Mermin and Wagner by means of an inequality proved by them. This inequality is a consequence of the well-known Bogolubov inequality (see [39]). The Dobrushin–Shlosman theorem can be found in [66].

3. Recently Briemont, Fontaine and Landau proved that under the conditions of the Dobrushin–Shlosman theorem the limit Gibbs distribution invariant under the action of the group G is unique (*Comm. Math. Phys.* **56** (1977), 281–296).

4. The Fröhlich–Simon–Spencer theorem was proved in their paper [64]. A more detailed discussion with new applications of the basic method is contained in a series of articles of Fröhlich [65–67].

CHAPTER IV

PHASE TRANSITIONS OF THE SECOND KIND AND THE RENORMALIZATION GROUP METHOD

1. Introduction

For large values of β the results of Chapter II give an idea of the structure of the phase diagram in classical lattice models, i.e. of the structure of the set of translation-invariant or periodic limit Gibbs distributions corresponding to Hamiltonians $\beta\mathscr{H}$. On the other hand, for small values of β the limit Gibbs distribution corresponding to the Hamiltonian $\beta\mathscr{H}$ is unique under sufficiently general conditions. It is clear that by varying β critical values $\beta=\beta_{cr}$ appear in the neighbourhood of which the structure of the set of translation-invariant or periodic limit Gibbs distributions changes. As examples we shall consider the ferromagnetic models, i.e. models having two translation-invariant or periodic ground states satisfying Peierls's condition which turn into each other if we change the sign of every variable. In this case it is natural to introduce the following definition.

Definition 4.1. *A number β_{cr} is called a ferromagnetic critical point iff in some neighbourhood of it for every $\beta<\beta_{cr}$ the translation-invariant limit Gibbs distribution is unique and for every $\beta>\beta_{cr}$ there exist exactly two translation-invariant limit Gibbs distributions.*

One could introduce analogous definitions for antiferromagnetic critical points or critical points connected with non-periodic ground states but we shall not do so since there are no exact results in this direction at the moment.

Examples such as the two-dimensional Ising model show that for β_{cr} the limit Gibbs distribution corresponding to the Hamiltonian $\beta_{cr}\mathscr{H}$ is unique, but the random variables $\varphi(x)$ cannot be regarded in any sense as the weakly dependent random variables studied in the classical probability theory. In the physics literature the following assumption is widespread: the joint distribution of similar random variables satisfies certain conditions of scale invariance or self-similarity. This assumption is of much more general interest since it shows that the breakdown of continuity or smoothness of morphological processes is often accompanied by the appearance of quantities fulfilling the scaling hypothesis.

We are going to introduce some parameters characterizing the structure of the set of translation-invariant limit Gibbs distributions in the neighbourhood of a ferromagnetic critical point β_{cr}.

1. Let P_0 be a translation-invariant limit Gibbs distribution for $\beta=\beta_{cr}$ such that $E\varphi(x)=0$,

$$E\varphi(x)\cdot\varphi(y)\sim\frac{\text{const}}{\|x-y\|^{d-2+\eta}}\quad\text{for}\quad\|x-y\|\to\infty$$

where d is the dimension of the system. Our reason for writing the exponent in the above way will be explained later. The index η is the main parameter. Set $\alpha=1+$ $+\frac{2}{d}-\frac{\eta}{d}$. If $\alpha\geqq 1$, then for the cube V_L of side L

$$E\left(\sum_{x\in V_L}\varphi(x)\right)^2=\sum_{x,y\in V_L}E(\varphi(x)\cdot\varphi(y))\sim$$

$$\sim\text{const}\cdot\sum_{x,y\in V_L}\frac{1}{\|x-y\|^{d-2+\eta}}=\text{const}\cdot|V_L|^{1+2/d-\eta/d}=\text{const}\cdot|V_L|^{\alpha}.$$

Since we consider the case when $\sum_{x\in V_L}\varphi(x)$ takes bounded values, $\alpha\leqq 2$ must hold. The inequality $\alpha>1$ indicates the presence of strongly dependent random variables.

2. Let us consider the translation-invariant limit Gibbs distributions $P(\beta)$ for $\beta<\beta_{cr}$. Suppose, further, that $E(\varphi(x))=0$, $E(\sum_{x\in V_L}\varphi(x))^2\sim\sigma(\beta)|V_L|$. It is natural to assume that $\sigma(\beta)\uparrow\infty$ when $\beta\uparrow\beta_{cr}$ and this singularity has the order of a power, i.e. $\sigma(\beta)\sim\text{const}\cdot(\beta_{cr}-\beta)^{-\gamma}$ for $\beta\uparrow\beta_{cr}$. Then the index γ characterizes the "speed of appearence of the dependence" for $\beta\uparrow\beta_{cr}$.

3. According to Definition 4.1 there exist for $\beta>\beta_{cr}$ two different translation-invariant limit Gibbs distributions $P_1(\beta)$, $P_2(\beta)$. Their difference appears already in the means

$$a_1(\beta)=E_{P_1(\beta)}(\varphi(x)),\quad a_2(\beta)=E_{P_2(\beta)}(\varphi(x)),$$

i.e.

$$a_1(\beta)\neq a_2(\beta).$$

Then $a_1(\beta)-a_2(\beta)\to 0$ for $\beta\downarrow\beta_{cr}$, and it is again natural to assume $|a_1(\beta)-a_2(\beta)|\sim$ $\sim\text{const}\cdot(\beta-\beta_{cr})^{\omega}$ for $\beta\downarrow\beta_{cr}$.

The quantities η,γ,ω are called critical indices. Besides the above-mentioned critical indices, other critical indices are used in the physics literature to characterize the statistical physical properties of a system in the neighbourhood of β_{cr}. The ultimate aim of the theory of critical points is to determine the critical indices corresponding to a given Hamiltonian. This problem is still far from being solved. In this chapter we shall treat some mathematical results concerning this problem.

2. Dyson's hierarchical models

In Dyson's work [59] devoted to the theory of phase transitions for one-dimensional models with long-range interaction auxiliary models were constructed which the author called hierarchical models. Later it turned out that hierarchical models are also of interest in the theory of critical points. In this section we give the definition of Dyson's hierarchical model, and in the next two sections we obtain a number of results on the critical point for these models.

Hierarchical models, or more precisely their Hamiltonians, are constructed inductively. At the nth step we consider a volume $V^{(n)}$, $|V^{(n)}|=2^n$ consisting of two subvolumes $V_1^{(n-1)}$, $V_2^{(n-2)}$, $|V_1^{(n-1)}|=|V_2^{(n-1)}|=2^{n-1}$. What we call a configuration φ in the volume $V^{(n)}$ is a function $\varphi(V^{(n)})=\{\varphi(s), s\in V^{(n)}\}$ defined on $V^{(n)}$ and taking the values ± 1. The Hamiltonian $\mathscr{H}_n(\varphi(V^{(n)}))$ is defined by a parameter c, $1<c<2$, and has the form

$$\mathscr{H}_n(\varphi(V^{(n)})) = \mathscr{H}_{n-1}(\varphi(V_1^{(n-1)}))+\mathscr{H}_{n-1}(\varphi(V_2^{(n-1)}))- -\frac{c^n}{2^{2n}}\left(\sum_{s\in V^{(n)}}\varphi(s)\right)^2. \tag{4.1}$$

Thus if we consider $n\geqq n_0$, then for the complete construction of the Hamiltonian we must give "initial conditions" in the form of an initial or priming Hamiltonian $\mathscr{H}_{n_0}(\varphi(V^{(n_0)}))$. For every $n>n_0$ the volume $V^{(n)}$ consists of two equal subvolumes $V_{i_1}^{(n-1)}$ $(i_1=1, 2)$; both again consist of two equal subvolumes $V_{i_1,i_2}^{(n-2)}$, etc., and this decomposition process can be carried out $n-n_0$ times.

Let $\xi^{(m)}$ denote the partition of $V^{(n)}$ into subvolumes $V_{i_1,i_2,\dots,i_{n-m}}^{(m)}$, and let $G_{n_0}^n$ denote the subgroup of the group of those permutations of the space $V^{(n)}|\xi^{(n-n_0)}$ which preserve every partition

$$\xi^{(m)}, \quad m = n-n_0, \quad n-n_0+1, \dots, n.$$

The last statement means that a permutation $g\in G_{n_0}^n$ moves an element of a partition $\xi^{(m)}$ into an element of the same partition. It is clear that the Hamiltonian $\mathscr{H}_n$ is invariant with respect to the group $G_{n_0}^n$, i.e.

$$\mathscr{H}_n(\varphi(V^{(n)})) = \mathscr{H}_n(g\varphi(V^{(n)})), \quad g\in G_{n_0}^n.$$

The term $-c^n 2^{-2n}(\sum_{s\in V^{(n)}}\varphi(s))^2$ in (4.1) describes the interaction between the configurations $\varphi(V_1^{(n-1)})$ and $\varphi(V_2^{(n-1)})$. It is important that this interaction is binary and depends on the values of the sums $\sum_{s\in V_1^{(n-1)}}\varphi(s)$ and $\sum_{s\in V_2^{(n-1)}}\varphi(s)$ only. This fact will be absolutely essential. The minus sign before this term means that we are dealing with a ferromagnetic model. The parameter c takes its values in the interval $1<c<2$ since for $c<1$ the interaction $c^n 2^{-2n}(\sum_{s\in V^{(n)}}\varphi(s))^2$ tends to 0 as $n\to\infty$ uniformly for all configurations, while for $c>2$ the energy of the configuration $\{\varphi(s)\equiv 1\}$

grows faster than the volume and the thermodynamical limit does not exist. The boundary cases $c=1, 2$ require a special analysis.

Let us assume that $n_0=1$. For any pair of points $s_1, s_2 \in V^{(n)}$ let us take the largest k for which the points s_1 and s_2 belong to different elements of the partition $\xi^{(k)}$, and set $d(s_1, s_2)=2^{n-k}$. In particular, for $s_1 \in V_1^{(n-1)}$, $s_2 \in V_2^{(n-1)}$, $k=1$, $d(s_1, s_2)=2^{n-1}$. For such s_1, s_2 the term $-c^n 2^{-2n}\varphi(s_1)\varphi(s_2)$, describing the interaction of $\varphi(s_1)$ and $\varphi(s_2)$ is equal to $-\frac{c}{4}\frac{\varphi(s_1)\varphi(s_2)}{d^\alpha(s_1, s_2)}$, $\alpha=2-\log_2 c$. It is clear that $1<\alpha<2$. Analogously we can consider any pair of points s_1, s_2. As a result we get that the Hamiltonian $\mathscr{H}_n(\varphi(V^{(n)}))$ can be written as

$$\mathscr{H}_n(\varphi(V^{(n)})) = \sum\nolimits_{i_1, i_2, \dots, i_{n-n_0}} \mathscr{H}_{n_0}(\varphi(V^{(n_0)}_{i_1, i_2, \dots, i_{n-n_0}})) -$$

$$- \sum_{\substack{s_1 \in V^{(n_0)}_{i_1, \dots, i_{n-n_0}} \\ s_2 \in V^{(n_0)}_{j_1, \dots, j_{n-n_0}} \\ (i_1, \dots, i_{n-n_0}) \neq (j_1, \dots, j_{n-n_0})}} \frac{a^{(n)}_{n_0}(s_1, s_2)}{d^\alpha(s_1, s_2)} \varphi(s_1)\varphi(s_2)$$

where the coefficients $a^{(n)}_{n_0}(s_1, s_2)$ are bounded from above and below by constants independent of n, n_0. Let us introduce the quantity $f_n(t; \beta)$ equal to the probability of the event that the sum $\Phi(\varphi(V^{(n)}))=\sum_{s \in V^{(n)}}\varphi(s)$ takes the value t, i.e.

$$f_n(t; \beta) = \frac{\sum_{\varphi(V^{(n)}): \Phi(\varphi(V^{(n)}))=t} e^{-\beta\mathscr{H}_n(\varphi(V^{(n)}))}}{\sum_{\varphi(V^{(n)})} e^{-\beta\mathscr{H}_n(\varphi(V^{(n)}))}} =$$

$$= \frac{1}{\Theta_n(\beta)} \sum\nolimits_{\varphi(V^{(n)}): \Phi(\varphi(V^{(n)}))=t} e^{-\beta\mathscr{H}_n(\varphi(V^{(n)}))}.$$

Then using the above expression for $\mathscr{H}_n$ the following sequence of equations can be derived

$$f_n(t; \beta) = \frac{\sum_{\Phi(\varphi(V^{(n)}))=t} e^{-\beta\mathscr{H}_n(\varphi(V^{(n)}))}}{\Theta_n(\beta)} =$$

$$= \frac{1}{\Theta_n(\beta)} \sum\nolimits_{t_1} \sum_{\substack{\Phi(\varphi(V_1^{(n-1)}))=t_1, \\ \Phi(\varphi(V_2^{(n-1)}))=t-t_1}} e^{-\beta\mathscr{H}_n(\varphi(V^{(n)}))} =$$

$$= \frac{1}{\Theta_n(\beta)} \sum\nolimits_{t_1} \sum_{\substack{\Phi(\varphi(V_1^{(n-1)}))=t_1, \\ \Phi(\varphi(V_2^{(n-1)}))=t-t_1}} e^{-\beta\mathscr{H}_{n-1}(\varphi(V_1^{(n-1)}))-\beta\mathscr{H}_{n-1}(\varphi(V_2^{(n-1)}))+\beta c^n 2^{-2n} t^2} =$$

$$= \frac{\Theta^2_{n-1}(\beta)}{\Theta_n(\beta)} e^{\beta c^n 2^{-2n} t^2} \sum\nolimits_{t_1} \sum_{\substack{\Phi(\varphi(V_1^{(n-1)}))=t_1, \\ \Phi(\varphi(V_2^{(n-2)}))=t-t_1}} \frac{e^{-\beta\mathscr{H}_{n-1}(\varphi(V_1^{(n-1)}))-\beta\mathscr{H}_{n-1}(\varphi(V_2^{(n-1)}))}}{\Theta^2_{n-1}} =$$

$$= \frac{\Theta^2_{n-1}(\beta)}{\Theta_n(\beta)} e^{\beta c^n 2^{-2n} t^2} \sum\nolimits_{t_1} f_{n-1}(t_1; \beta) f_{n-1}(t-t_1; \beta).$$

The last relation shows that the distribution f_n can be expressed recursively by the distribution f_{n-1}. The factor $\Theta_{n-1}^2(\beta)/\Theta_n(\beta)$ is determined by the normalization condition. Now we can forget about the Hamiltonian $\mathscr{H}_n$ and deal only with the sequence of recursive relations:

$$f_n(t;\,\beta) = K_n(\beta)\,e^{\beta c^n 2^{-2n} t^2} \sum_{t_1=-2^{n-1}}^{2^{n-1}} f_{n-1}(t_1;\,\beta) f_{n-1}(t-t_1;\,\beta) \tag{4.2}$$

$$K_n^{-1}(\beta) = \sum_{t=-2^n}^{2^n} e^{\beta c^n 2^{-2n} t^2} \sum_{t_1=-2^{n-1}}^{2^{n-1}} f_{n-1}(t_1;\,\beta) f_{n-1}(t-t_1;\,\beta) \tag{4.3}$$

t and t_1 are even numbers.

Our aim is to investigate the behaviour of the hierarchical models in the neighbourhood of β_{cr}. As for $\beta<\beta_{cr}$, we expect the typical values of the total spin $\Phi(\varphi(V^{(n)}))$ to be of the order $t\sim 2^{n/2}$. For such values of t the factor $\exp(\beta c^n 2^{-2n} t^2)\to 1$ when $n\to\infty$, and therefore the sequence $f_n(t;\beta)$ behaves like a sequence of usual convolutions, asymptotically. But for $\beta>\beta_{cr}$ the typical values of t are of the order 2^n and the factor $\exp(\beta c^n 2^{-2n} t^2)$ is dominant. Let us now take the following assumption, to be justified by the analysis which follows: the point $\beta=\beta_{cr}$ is characterized by the relation $c^n 2^{-2n} t^2 \sim 1$ for typical values of t. If this assumption is true, then it is natural to set $t=z2^n c^{-n/2}$ and suppose that $z\sim 1$ for $\beta=\beta_{cr}$. Further, let

$$g_n(z;\,\beta) = f_n(z2^n c^{-n/2};\,\beta)2^n c^{-n/2}.$$

Then putting the last expression into (4.2) we get

$$g_n(z;\,\beta) = L_n(\beta)\,e^{\beta z^2} \sum_{\frac{z_1+z_2}{2}=\frac{z}{\sqrt{c}}} g_{n-1}(z_1;\,\beta)\, g_{n-1}(z_2;\,\beta)\, c^{n/2} 2^{-n} \tag{4.4}$$

$$\left(t = 2j\cdot\frac{c^{n/2}}{2^n},\quad \text{where}\quad -2^n \leqq j \leqq 2^n\right).$$

The factor $L_n(\beta)$ is, as above, determined by the normalization condition. If our assumption is true, then we have a good reason to expect that for $n\to\infty$ the functions $g_n(z;\beta)$ tend to a limit, and that the limit function $g(z;\beta)$ satisfies the equation

$$g(z;\,\beta) = L(\beta)\,e^{\beta z^2} \int_{-\infty}^{\infty} g\left(\frac{z}{\sqrt{c}}+u;\,\beta\right) g\left(\frac{z}{\sqrt{c}}-u;\,\beta\right) du. \tag{4.5}$$

$$L^{-1}(\beta) = \int_{-\infty}^{\infty} e^{\beta z^2}\,dz \int_{-\infty}^{\infty} g\left(\frac{z}{\sqrt{c}}+u;\,\beta\right) g\left(\frac{z}{\sqrt{c}}-u;\,\beta\right) du. \tag{4.5'}$$

Therefore the problem is to find the positive integrable solutions of the nonlinear integral equation

$$g(z;\,\beta) = e^{\beta z^2} \int_{-\infty}^{\infty} g\left(\frac{z}{\sqrt{c}}+u;\,\beta\right) g\left(\frac{z}{\sqrt{c}}-u;\,\beta\right) du, \tag{4.6}$$

since from the solutions of this equation we can obtain the ones of interest to us by normalization. Let us also note, that if $g(z;\beta_0)$ is a solution of Equation (4.5) for some β_0, then the solution of (4.6) for any β can be obtained from the relation

$$g(z;\beta)=g\left(z\sqrt{\beta/\beta_0};\beta_0\right)\sqrt{\beta/\beta_0}.$$

Thus it is natural to consider one-parameter families of functions $\{g(z;\beta)\}$ satisfying Equation (4.6) for every β, $0<\beta<\infty$.

Now we can formulate the problem of the critical points for hierarchical models more precisely. Let $\{g(z;\beta)\}$ be a family of the solutions of Equations (4.5) and (4.5′). We have to describe the set $\mathscr{U}$ of initial Hamiltonians $\mathscr{H}_{n_0}(\varphi(V^{(n_0)}))$ possessing the following property: for every $\mathscr{H}_{n_0}\in\mathscr{U}$ there exists a $\beta_{\mathrm{cr}}(\mathscr{H}_{n_0})$ such that

1) the probability distribution defined by the function $g_n(z,\beta_{\mathrm{cr}})$ tends weakly to the probability distribution with the density $g(z;\beta_{\mathrm{cr}})$ as $n\to\infty$;
2) if $\beta<\beta_{\mathrm{cr}}$, then there exists a function $\sigma(\beta)$ such that for $n\to\infty$

$$f_n(t;\beta)\sim\frac{1}{\sqrt{2\pi\sigma(\beta)2^n}}e^{-t^2/2\sigma(\beta)2^n}$$

whenever t satisfies the inequality $|t2^{-n/2}|\leqq A$ for every A independent of n, and $\sigma(\beta)\sim\mathrm{const}\cdot(\beta_{\mathrm{cr}}-\beta)^{-\gamma}$ for $\beta\uparrow\beta_{\mathrm{cr}}$; in addition the critical index γ does not depend on $\mathscr{H}_{n_0}$, and is determined only by the solution $g(z;\beta)$;

3) for $\beta>\beta_{\mathrm{cr}}$ there exists a function $m(\beta)$ such that the mean spin $\frac{1}{2^n}\Phi(\varphi(V^{(n)}))$ tends to $\pm m(\beta)$ for $n\to\infty$ in probability generated by the Gibbs distribution defined by the Hamiltonian $\beta\mathscr{H}_n$,

i.e.
$$\lim_{n\to\infty}P\left\{\frac{1}{2^n}\Phi(\varphi(V^{(n)}))<-m(\beta)\right\}=0,$$

$$\lim_{n\to\infty}P\left\{\frac{1}{2^n}\Phi(\varphi(V^{(n)}))<m(\beta)\right\}=\frac{1}{2},$$

$$\lim_{n\to\infty}P\left\{\frac{1}{2^n}\Phi(\varphi(V^{(n)}))>m(\beta)\right\}=1;$$

in addition $m(\beta)\sim\mathrm{const}\cdot|\beta-\beta_{\mathrm{cr}}|^{\omega}$, where the number ω does not depend on $\mathscr{H}_{n_0}$ but it is solely determined by the solution $g(z;\beta)$.*

In this connection it is natural to introduce the following definition.

* Properties 2), 3) were formulated in the most natural way from the point of view of probability theory.

Definition 4.2. *The solution $g(z; \beta)$ of Equation (4.6) is said to be thermodynamically stable iff there exists a natural number n_0 such that the set $\mathcal{U}$ described above is open (with respect to the natural topology* for Hamiltonians satisfying the symmetry condition).*

Now we describe how the solution of the problem stated above is conventionally approached in the physics literature. We shall consider (4.4) as a nonlinear transformation $T(\beta)$ transforming $g_{n-1}(z; \beta)$ to $g_n(z; \beta)$. Let us consider the curve $g = \{g(z; \beta)\}$ of the solutions of (4.5) as a curve parametrized by β.

Every point on this curve is a fixed point of the nonlinear transformation $T(\beta)$: $: g(z; \beta) = T(\beta) g(z; \beta)$. As usual in the theory of nonlinear mappings, we investigate the stability of $g(z; \beta)$ in the linear approximation and we assume that the tangent space at the point $g(z; \beta)$ consists of a one-dimensional unstable subspace and a stable subspace of co-dimension 1. It follows from this fact in the finite-dimensional theory that through every point $g(z; \beta)$ there passes the stable submanifold $\Gamma^{(s)}_{(\beta)}$ of co-dimension 1 belonging to this point, consisting of such g' that $T^n(\beta) g' \to g(z; \beta)$ when $n \to \infty$. The union of these submanifolds for all β will form a neighbourhood of the curve g.

Let us consider the curve of initial distributions $g_{n_0} = \{g_{n_0}(z; \beta)\}$. We can see from the geometrical picture (Figure 4.1) that β_{cr} is determined by the relation $g_{n_0}(z; \beta_{cr}) \in \Gamma^{(s)}(\beta_{cr})$. The experience from the theory of nonlinear transformations shows that

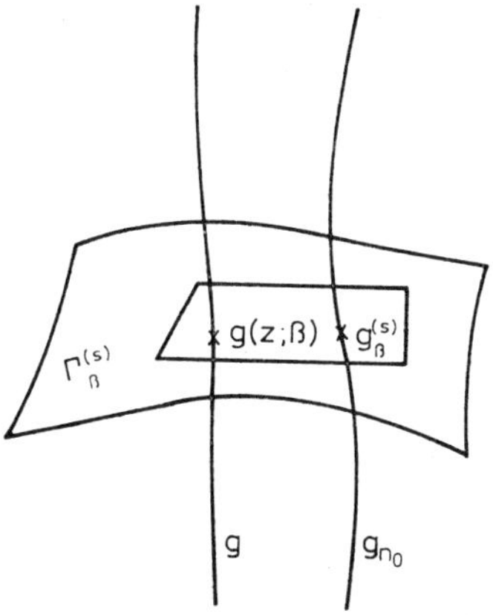

FIGURE 4.1

* Let us define more precisely the topology which we have in mind. The space of all possible initial Hamiltonians $\mathcal{H}(\varphi(V_{n_0}))$ is a vector space of dimension 2^{n_0} with the usual topology. The space of symmetric Hamiltonians $\mathcal{H}(\varphi(V_{n_0})) = \mathcal{H}(-\varphi(V_{n_0}))$ is a closed subset of this space. Our topology is the induced topology on this subset.

the stable manifolds $\Gamma^{(s)}(\beta_{cr})$ cannot be constructed explicitly since they are the solutions of complicated functional equations which are constructed by means of successive approximations only resulting in an existence proof of such manifolds. Usually we know the tangent space $\tau^{(s)}(\beta)$ to $\Gamma^{(s)}(\beta)$ at the point $g(z;\beta)$ generated by the stable eigenvectors of the linearized problem, and this is sufficient for the majority of the problems. Also, the union of such tangent subspaces covers a neighbourhood of the curve g.

Let us assume that n_0 is large and the initial Hamiltonian is chosen in such a way that the curve $g_{n_0}(z;\beta)$ is sufficiently close to the curve g and we can consider the function $\bar{\beta}=\lambda(\beta)$ defined by the relation $g_{n_0}(z;\beta)\in\tau^{(s)}(\bar{\beta})$. If we assume that $\tau^{(s)}(\beta)$ gives an approximate representation of $\Gamma^{(s)}(\beta)$, then the solution of the equation $\beta=\lambda(\beta)$ is an approximate value for β_{cr}. If the solution $g(z;\beta)$ is thermodynamically stable, then the real value of β_{cr} lies in the neighbourhood of this approximate value.

The argument presented above can be regarded as heuristic considerations, leading to the following programme of action:

1) having solutions $g(z;\beta)$, investigate their stability in the linear approximation and find those among them which have only one unstable direction;
2) construct a Hamiltonian $\mathscr{H}_{n_0}$ such that the function $\lambda(\beta)$ corresponding to it has a transversal intersection with the diagonal $\bar{\beta}=\beta$;
3) prove that there exists a point β_{cr} close to the solution of the equation $\beta=\lambda(\beta)$ for which $g_n(z;\beta_{cr})\to g(z;\beta)$;
4) investigate the behaviour of $g_n(z;\beta)$ for β close to β_{cr}.

It will be shown later how this programme can be carried out for some special solutions of (4.6).

3. Gaussian solutions

For every c, $1<c<2$, Equations (4.5), (4.5′) have a Gaussian solution

$$g^{(0)}(z;\beta)=\sqrt{\frac{a_0(\beta)}{\pi}}\,e^{-a_0(\beta)z^2},\quad a_0(\beta)=\frac{\beta c}{2-c}.$$

Corresponding to the above programme we have to begin our investigation with the analysis of the stability of this solution. It is a bit simpler to deal with Equation (4.6). Putting $g(z;\beta)=e^{-a_0(\beta)z^2}h(z;\beta)$ in (4.6) we get

$$h(z;\beta)=\int_{-\infty}^{\infty}e^{-2a_0(\beta)u^2}\,h\left(\frac{z}{\sqrt{c}}+u;\beta\right)h\left(\frac{z}{\sqrt{c}}-u;\beta\right)du=Q(\beta)\,h. \tag{4.7}$$

Solution $h(z;\beta)=C(\beta)=\sqrt{\dfrac{2a_0(\beta)}{\pi}}$ of (4.7) corresponds to the Gaussian solution.

Since the solutions of (4.7) for different β are connected by the relation

$$h(z;\,\beta)=h\left(z\sqrt{\frac{\beta}{\beta_0}};\,\beta_0\right)\sqrt{\frac{\beta}{\beta_0}},$$

it is sufficient to consider β_0 with the property $2a_0(\beta_0)=\frac{1}{2}$. For this reason the argument β_0 will be omitted later on. To investigate the property of stability let us put $h(z)=h_0(z)+\varepsilon G(z)=\frac{1}{\sqrt{2\pi}}+\varepsilon G(z)$ and write

$$Q\big(h_0(z)+\varepsilon G(z)\big)=Q\big(h_0(z)\big)+\varepsilon A\big(G(z)\big)+\ldots$$

where dots mean terms of higher order in ε. The linear operator A is the differential of the nonlinear transformation $Q(\beta_0)$. In our case A is the well-known Gauss integral operator (Bateman and Erdélyi, *Higher transcendental functions,* McGraw-Hill, New York, 1955):

$$A(G)\,z=\frac{2}{\sqrt{2\pi}}\int_{-\infty}^{\infty}e^{-u^2/2}\,G\left(\frac{z}{\sqrt{c}}+u\right)du. \tag{4.8}$$

Also, the operator $A(\beta)$ defined for any β by the equality

$$A(\beta)(G)(z)=2\sqrt{\frac{2a_0(\beta)}{\pi}}\int_{-\infty}^{\infty}e^{-2a_0(\beta)\,u^2}\,G\left(\frac{z}{\sqrt{c}}+u\right)du \tag{4.8$'$}$$

is needed.

It is clear that if $G(z)=z^n+\ldots$ is a polynomial of degree n, then $A(G)$ is a polynomial of degree n as well in which the coefficient of z^n is equal to $2c^{-n/2}$. Therefore for every n, we can construct a polynomial G_n of degree n which is an eigenvector of the operator A with eigenvalue $\lambda_n=2c^{-n/2}$. We show that G_n is the nth Hermite polynomial corresponding to the Gaussian density function $\exp\left\{-\frac{c-1}{2c}z^2\right\}$. Indeed let us observe that the operator A is a symmetric operator on the Hilbert space $L^2\left(R^1,\exp\left(-\frac{c-1}{2c}z^2\right)\right)$. Indeed, for any G' and G'' we have

$$\int_{-\infty}^{\infty}(AG')G''e^{-((c-1)z^2)/2c}\,dz=$$

$$=\frac{2}{\sqrt{2\pi}}\iint_{-\infty}^{\infty}e^{-\frac{u^2}{2}-\frac{c-1}{2c}z^2}\,G'\left(\frac{z}{\sqrt{c}}+u\right)G''(z)\,dz\,du=$$

$$=\iint_{-\infty}^{\infty}G'(w)\,G''(z)\,e^{-w^2/2+(wz/\sqrt{c})-z^2/2}\,dw\,dz=\int_{-\infty}^{\infty}A(G'')G'e^{-((c-1)z^2/2c)}\,dz.$$

Therefore the polynomials G_n are orthogonal with respect to the density function $\exp\left\{-\frac{c-1}{2c}z^2\right\}$; hence by definition they are the Hermite polynomials.

Also, it follows from the above consideration that the eigenfunctions G_n form a complete orthogonal system in the space $L^2\left(R^1, \exp\left(-\frac{c-1}{2c}z^2\right)\right)$. Let us observe that in our problems we are dealing with even Hamiltonians (i.e. Hamiltonians invariant under the $\pm$ symmetry $\varphi(V^{(n)}) \to -\varphi(V^{(n)})$). Therefore we have to investigate the spectrum of the operator A in the space of even functions only; thus we are interested only in even n ($n=2n'$, $n'=0, 1, \ldots$). Finally we are only interested in those perturbations G for which $h_0+\varepsilon G$ differs from a probability density function in higher order terms of ε when ε is small. This means that we have to drop the constants and consider the spectrum of the operator A in the subspace of all even functions whose integral is equal to zero, i.e. the numbers $\lambda_2, \lambda_4, \lambda_6 \ldots$ are to be considered. Let us observe that $\lambda_2 = \frac{2}{c} > 1$ for every c, $1<c<2$. Further $\lambda_4 = 2/c^2 > \lambda_6 > \ldots$. Returning to the general picture described above, we expect that the Gaussian solution is thermodynamically stable if $\lambda_4 \leqq 1$ i.e. $c \geqq \sqrt{2}$. Indeed the following theorem is true.

Theorem 4.1. *Let $\beta_{cr}^{(0)}>0$, $\varepsilon>0$ be fixed and $\sqrt{2}<c<2$. Then an n_0 can be found as well as an open set $\mathcal{U}$ in the space of even Hamiltonians $\mathcal{H}_{n_0}(\varphi(V_{n_0}))$ with the property that for every $\mathcal{H}_{n_0} \in \mathcal{U}$ there exists a $\beta_{cr} = \beta_{cr}(\mathcal{H}_{n_0})$ $(|\beta_{cr}-\beta_{cr}^{(0)}|<\varepsilon)$ such that the probability distribution function corresponding to the function $g_n(z; \beta_{cr})$ converges weakly to the Gaussian distribution for $n \to \infty$.*

The proof of the theorem can conveniently be carried out by looking at the partition functions

$$Z_n(t; \beta) = f_n(t; \beta)\Theta_n(\beta), \quad \Theta_n(\beta) = \sum_{\varphi(V^{(n)})} e^{-\beta\mathcal{H}_n(\varphi(V^{(n)}))}.$$

It follows from the recurrence equations for $f_n(t; \beta)$ that

$$Z_{n+1}(t; \beta) = e^{\beta c^{n+1} 2^{-2(n+1)} t^2} \sum_{t_1+t_2=t} Z_n(t_1; \beta) Z_n(t_2; \beta). \tag{4.9}$$

Here t_1, t_2 are even integer numbers, $|t_1|, |t_2| \leqq 2^n$, $|t| \leqq 2^{n+1}$. The induction hypotheses on the structure of $Z_n(t; \beta)$ are in the centre of our considerations. We are going to prove that their validity at the $(n+1)$st step follows from their validity at the nth step. Then we prove that they are fulfilled for some $n=n_0$. The point β for which the induction hypotheses are fulfilled for every $n \geqq n_0$ will be critical.

The induction hypotheses

Set $D^{(n)}=\{t:|t|\leqq D\sqrt{n}\,c^{-n/2}2^n\}$ where $D=D(c)$ is a constant and let $N=N(c)$ be an integer; both will be chosen sufficiently large. Instead of the variable t, at the nth step it is more convenient to use the variable z, $z=c^{n/2}2^{-n}t$, obtained by normalization at the critical point (see above). The interval $D^{(n)}$ expressed by the variable z has the form:

$$D^{(n)}=\{z:\ |z|\leqq D\sqrt{n}\}.$$

Let us assume that at the nth step there is given an interval $\beta^{(n)}=[\beta_-^{(n)},\beta_+^{(n)}]$ of the values of parameter β, for which the following informative hypotheses hold:

$\mathrm{Ind}_1^{(n)})$. If $\beta\in\beta^{(n)}$ and $z\in D^{(n)}$, then,

$$Z_n(t;\beta)=L_n(\beta)c^{n/2}2^{-n}\exp\{-a_0(\beta)z^2-B_2^{(n)}(\beta)(2c^{-1})^n\cdot \tag{4.10}$$
$$\cdot G_2(z;\beta)-B_4^{(n)}(\beta)(2c^{-2})^nG_4(z;\beta)-$$
$$-\textstyle\sum_{k=3}^N B_{2k}(\beta)(2c^{-2})^n\lambda_1^n G_{2k}(z;\beta)-Q^{(n)}(z;\beta)-R^{(n)}(z;\beta)\}.$$

Here $L_n(\beta)$ is a constant depending only on β and n, $G_{2k}(z;\beta)$ are the Hermite polynomials corresponding to the Gaussian density function $\sqrt{\dfrac{\beta\gamma}{\pi}}e^{-\beta\gamma z^2}$, $\gamma=\dfrac{2a_0(\beta)(c-1)}{\beta c}$, λ_1 is a constant, $0<\lambda_1<1$; the function $Q^{(n)}(z;\beta)$, defined for all real z, is even in z, is also bounded, and

$$\int Q^{(n)}(z;\beta)G_{2k}(z;\beta)e^{-\beta\gamma z^2}dz=0,\quad k=0,1,\ldots,N$$

and $|Q^{(n)}(z;\beta)|\leqq\lambda_1^n(2c^{-2})^n$ for $z\in D^{(n)}$; the function $R^{(n)}(z;\beta)$ is defined for $z=c^{n/2}2^{-n}t\in D^{(n)}$ (t is integer, $|t|\leqq 2^n$) and $|R^{(n)}(z;\beta)|\leqq\lambda_1^{3n/2}(2c^{-2})^n$.

Further, it is assumed that λ_1 is sufficiently close to 1.

$\mathrm{Ind}_2^{(n)})$. $B_2^{(n)}(\beta)$ is a continuous function of β,

$$B_2^{(n)}(\beta_-^{(n)})=\lambda_1^nc^{-n},\quad B_2^{(n)}(\beta_+^{(n)})=-\lambda_1^nc^{-n}$$

and $|B_2^{(n)}(\beta)|\leqq\lambda_1^nc^{-n}$ for every $\beta\in\beta^{(n)}$.

$\mathrm{Ind}_3^{(n)})$. For every $t\notin D^{(n)}$.

$$Z_n(t;\beta)\leqq L_n(\beta)c^{n/2}2^{-n}\exp\{-a_0(\beta)c^n2^{-2n}t^2-\bar{B}_4^{(n)}(\beta)2^{-3n}t^4\}=$$
$$=L_n(\beta)c^{n/2}2^{-n}\exp\{-a_0(\beta)z^2-\bar{B}_4^{(n)}(\beta)(2c^{-2})^nz^4\};$$

$$\bar{B}_4^{(n)}(\beta)\text{ is a constant; }\bar{B}_4^{(n)}(\beta)\leqq\frac{1}{2}\,B_4^{(n)}(\beta).$$

$\mathrm{Ind}_4^{(n)})$. The quantities L_n, $B_{2k}^{(n)}$, $\bar{B}_4^{(n)}$ depend continuously on β; $Q^{(n)}$ and $R^{(n)}$ depend continuously on β for every z.

The basic lemma. *Let* $\mathrm{Ind}_1^{(n)}$–$\mathrm{Ind}_4^{(n)}$ *be fulfilled. Then there exists an interval* $\beta^{(n+1)}\subset\beta^{(n)}$ *for which the partition functions* $Z_{n+1}(t;\beta)$ *satisfy* $\mathrm{Ind}_1^{(n+1)}$–$\mathrm{Ind}_4^{(n+1)}$.

Moreover one can find a constant $\lambda_2<1$, such that

$$|B_{2k}^{(n)}-B_{2k}^{(n+1)}|\leqq\lambda_2^n,\quad k=2,$$
$$|B_{2k}^{(n+1)}-c^{-k+2}\lambda_1^{-1}B_{2k}^{(n)}|\leqq\lambda_2,\quad k=3,\ldots,N.$$
$$Q^{(n+1)}(z;\beta)=A(\beta)(Q^{(n)}(z;\beta))+Q_1^{(n)}(z;\beta),$$

where $|Q_1^{(n)}(z;\beta)|\leqq\text{const}\cdot n^{2N}\lambda_1^{3n/2}(2c^{-2})^n$, const *does not depend on n, while the operator $A(\beta)$ is defined by (4.8′). The constant $\bar{B}_4^{(n+1)}$ is equal to $\bar{B}_4^{(n)}(1-n^N\lambda_1^{3n/2})$.*

Proof. Let $z\in D^{(n+1)}$. We have for $t=z2^{n+1}c^{-(n+1)/2}$

$$Z_{n+1}(t;\beta)=e^{\beta c^{n+1}2^{-2(n+1)}t^2}\sum_{t_1+t_2=t}Z_n(t_1;\beta)Z_n(t_2;\beta)=$$
$$=e^{\beta z^2}\sum_{z_1+z_2=2z/\sqrt{c}}Z_n(z_12^nc^{-n/2};\beta)\cdot Z_n(z_22^nc^{-n/2};\beta)=$$
$$=e^{\beta z^2}\sum_u Z_n\left(\left(\frac{z}{\sqrt{c}}+u\right)2^nc^{-n/2};\beta\right)Z_n\left(\left(\frac{z}{\sqrt{c}}-u\right)2^nc^{-n/2};\beta\right),$$

where $2^{n-1}c^{-n/2}z_1$ and $2^{n-1}c^{-n/2}z_2$ are integer numbers. Here the step length of the variable u is $\Delta_n=c^{n/2}2^{-n}$ taking such values that $\left(\frac{2z}{\sqrt{c}}\pm u\right)2^nc^{-n/2}$ are even numbers, not exceeding 2^n in absolute value. Let us divide the last sum into two terms.

$$e^{\beta z^2}\sum_u Z_n\left(\left(\frac{z}{\sqrt{c}}+u\right)2^nc^{-n/2};\beta\right)Z_n\left(\left(\frac{z}{\sqrt{c}}-u\right)2^nc^{-n/2};\beta\right)=$$
$$=e^{\beta z^2}\sum_{|u|\leqq\frac{D}{6\sqrt{c}}\sqrt{n}}Z_n\left(\left(\frac{z}{\sqrt{c}}+u\right)2^nc^{-n/2};\beta\right)Z_n\left(\left(\frac{z}{\sqrt{c}}-u\right)2^nc^{-n/2};\beta\right)+$$
$$+e^{\beta z^2}\sum_{|u|>\frac{D}{6\sqrt{c}}\sqrt{n}}Z_n\left(\left(\frac{z}{\sqrt{c}}+u\right)2^nc^{-n/2};\beta\right)Z_n\left(\left(\frac{z}{\sqrt{c}}-u\right)2^nc^{-n/2};\beta\right)=\sum\nolimits_1+\sum\nolimits_2.$$

Let us first investigate $\sum_1$ noticing that, for the possible values of u, $z/\sqrt{c}\pm u\in D^{(n)}$. Therefore we can use $\text{Ind}_1^{(n)}$ and replace Z_n by its representation (4.10). Moreover in virtue of $\text{Ind}_1^{(n)}$, and $\text{Ind}_2^{(n)}$ we get

$$\begin{aligned}(4.11)\qquad &|B_2^{(n)}(\beta)(2c^{-1})^nG_2(z;\beta)|+|B_4^{(n)}(\beta)(2c^{-2})^nG_4(z;\beta)|+\\ &+\sum\nolimits_{k=3}^N|B_{2k}^{(n)}(\beta)|(2c^{-2})^n\lambda_1^n|G_{2k}(z;\beta)|+|Q^{(n)}|+\\ &+|R^{(n)}|\leqq\text{const}\cdot n^N(\max\nolimits_{\beta\in\beta_k^{(n)}}(\beta_2^{(n)}(2c^{-1})^n,|B_{2k}^{(n)}(\beta)|)(2c^{-2})^n),\end{aligned}$$

where const depends on N, but does not depend on n. Let us assume that the lemma is already proved for $n_0\leqq k\leqq n$ (n_0 is sufficiently large). Then all numbers $B_{2k}^{(n)}(\beta)$ are bounded in absolute value by a constant independent of n, k, β, which is denoted by B. Therefore (4.11)$\to0$ for $n\to\infty$ and owing to the statement of the lemma we can expand the corresponding expression in $\sum_1$ into a Taylor series up to the term of second order. We get

$$\sum_1 = L_n^2(\beta)\left(\frac{2}{\sqrt{c}}\right) c^{(n+1)/2}2^{-n-1}e^{\beta z^2-(2a_0(\beta)/c)z^2}\cdot$$

$$\cdot\sum_{|u|\leq\frac{D}{6\sqrt{c}}\sqrt{n}} e^{-2a_0(\beta)u^2}c^{n/2}2^{-n-1}\Bigg[1-B_2^{(n)}(\beta)(2c^{-1})^n\left[G_2\left(\frac{z}{\sqrt{c}}+u,\beta\right)+\right.$$

$$\left.+G_2\left(\frac{z}{\sqrt{c}}-u,\beta\right)\right]-B_4^{(n)}(\beta)(2c^{-2})^n\left[G_4\left(\frac{z}{\sqrt{c}}+u,\beta\right)+G_4\left(\frac{z}{\sqrt{c}}-u,\beta\right)\right]-$$

$$-\sum_{k=3}^N B_{2k}^{(n)}\lambda_1^n(2c^{-2})^n\left[G_{2k}\left(\frac{z}{\sqrt{c}}+u,\beta\right)+G_{2k}\left(\frac{z}{\sqrt{c}}-u,\beta\right)\right]-$$

$$-\left[Q^{(n)}\left(\frac{z}{\sqrt{c}}+u,\beta\right)+Q^{(n)}\left(\frac{z}{\sqrt{c}}-u,\beta\right)\right]-$$

$$-\left[R^{(n)}\left(\frac{z}{\sqrt{c}}+u,\beta\right)+R^{(n)}\left(\frac{z}{\sqrt{c}}-u,\beta\right)\right]+T^{(n)}\Bigg].$$

Here $T^{(n)}$ is the error. It is clear that $T^{(n)}$ has the order of the square of the whole expression expanded. Therefore we get from (4.11) the inequality

$$T^{(n)} \leqq \text{const}\cdot B^2(2c^{-2})^{2n}n^N. \tag{4.12}$$

The next step is the replacement of the sum by an integral. For this purpose we have to extend the functions $R^{(n)}$ and $T^{(n)}$ over all values of the argument. Let them be step functions, constant on every interval of length Δ_n. The error arising from the replacement of the sum by an integral will be called $r_1^{(n)}$. Now we can write

$$e^{a_0(\beta)z^2}\sum_1 = L_n^2(\beta)2^{-n-1}c^{(n+1)/2}c^{-1/2}2\sqrt{\frac{\pi}{2a_0(\beta)}}\cdot$$

$$\cdot\left\{\sqrt{\frac{2a_0(\beta)}{\pi}}\int_{|u|\leq\frac{D}{6\sqrt{c}}\sqrt{n}} e^{-2a_0(\beta)u^2}du\left[1-2\left(B_2^{(n)}(\beta)(2c^{-1})^nG_2\left(\frac{z}{\sqrt{c}}+u,\beta\right)-\right.\right.\right.$$

$$-B_4^{(n)}(\beta)(2c^{-2})^nG_4\left(\frac{z}{\sqrt{c}}+u,\beta\right)-$$

$$\left.-\sum_{k=3}^N B_{2k}^{(n)}\lambda_1^n(2c^{-2})^nG_{2k}\left(\frac{z}{\sqrt{c}}+u,\beta\right)\right)-$$

$$\left.\left.-Q^{(n)}\left(\frac{z}{\sqrt{c}}+u,\beta\right)-R^{(n)}\left(\frac{z}{\sqrt{c}}+u,\beta\right)+T^{(n)}\right]+r_1^{(n)}\right\}=$$

$$=L_n^2(\beta)2^{-n-1}c^{(n+1)/2}c^{-1/2}2\sqrt{\frac{\pi}{2a_0(\beta)}}\,(1-B_2^{(n)}(\beta)(2c^{-1})^nA(\beta)\,G_2-$$

$$-B_4^{(n)}(2c^{-2})^nA(\beta)\,G_4-\sum_{k=3}^N B_{2k}^{(n)}\lambda_1^n(2c^{-2})^nA(\beta)\,G_{2k}-$$

$$-A(\beta)Q^{(n)}-A(\beta)R^{(n)}+T_1^{(n)}+r_1^{(n)}+r_2^{(n)}).$$

Here $T_1^{(n)}$ is the integral of $T^{(n)}$. The error $r_2^{(n)}$ comes from the fact that in the definition of A the integral should be taken on the whole real axis, while in the last expression the integral is only taken over the interval $|u| \leqq \frac{D}{6\sqrt{c}} \sqrt{n}$. As a result we get

$$e^{a_0(\beta)z^2} \sum\nolimits_1 = L_n^2(\beta) 2^{-n-1} c^{(n+1)/2} c^{-1/2} 2 \sqrt{\frac{\pi}{2a_0(\beta)}} \cdot$$

$$\cdot \Big(1 - \frac{2}{c} B_2^{(n)} (2c^{-1})^n G_2(z;\beta) - \frac{2}{c^2} B_4^{(n)} (2c^{-2})^n G_4(z;\beta) -$$

$$- \sum\nolimits_{k=3}^{N} \frac{2}{c^k} B_{2k}^{(n)} \lambda_1^n (2c^{-2})^n G_{2k}(z;\beta) - A(\beta) Q^{(n)} -$$

$$- A(\beta) R^{(n)} + T_1^{(n)} + r_1^{(n)} + r_2^{(n)}\Big) =$$

$$= L_n^2(\beta) 2^{-n-1} c^{(n+1)/2} c^{-1/2} 2 \sqrt{\frac{\pi}{2a_0(\beta)}} \cdot$$

$$\cdot \exp\Big\{- B_2^{(n)} (2c^{-1})^{n+1} G_2(z;\beta) - B_4^{(n)} (2c^{-2})^{n+1} G_4(z;\beta) -$$

$$- \sum\nolimits_{k=3}^{N} B_{2k}^{(n)} \lambda_1^n 2^{n+1} c^{-2n-k} G_{2k}(z;\beta) - A(\beta) Q^{(n)} -$$

$$- A(\beta) R^{(n)} + T_1^{(n)} + r_1^{(n)} + r_2^{(n)} + T_2^{(n)}\Big\}.$$

The term $T_2^{(n)}$ comes from the difference between $1-\varepsilon$ and $e^{-\varepsilon}$, $A(\beta)R^{(n)}$ is an even function defined for $z \in D^{(n+1)}$. We can write:

$$A(\beta) R^{(n)} = \sum\nolimits_{k=0}^{N} b_{2k}^{(n)} G_{2k}(z;\beta) + \bar{Q}_1^{(n)}(z;\beta),$$

$Q_1^{(n)}(z;\beta) = \bar{Q}_1^{(n)}(z;\beta) \chi_{D^{(n+1)}}(z)$ where the coefficients $b_{2k}^{(n)}$ are chosen so that

$$\int_{-\infty}^{\infty} Q_1^{(n)}(z;\beta) G_{2k}(z;\beta) e^{-\beta\gamma z^2} dz = 0.$$

It follows from the inequalities $|R^{(n)}| \leqq \lambda_1^{3n/2} (2c^{-2})^n$ and $\max_{z \in D_{n+1}} |G_{2k}^{(z)}(z;\beta)| < $ $<\text{const} \cdot n^k$ that $|b_{2k}^{(n)}| \leqq \text{const} \cdot n^N \lambda_1^{3n/2} (2c^{-2})^n$ and

$$|\bar{Q}_1^{(n)}(z;\beta)| \leqq \text{const} \cdot n^N (2c^{-2})^n \lambda_1^{3n/2}. \tag{4.13}$$

Now we are in the position of formulating our basic recursive equations:

$$B_2^{(n+1)} = B_2^{(n)} + b_2^{(n)} (c2^{-1})^{n+1}$$

$$B_4^{(n+1)} = B_4^{(n)} + b_4^{(n)} (c^2 2^{-1})^{n+1}$$

$$B_{2k}^{(n+1)} = B_{2k}^{(n)} c^{-k+2} \lambda_1^{-1} + b_{2k}^{(n)} (c^2 2^{-1})^{n+1} \lambda_1^{-n-1}, \quad 3 \leqq k \leqq N.$$

$$Q^{(n+1)} = A(\beta) Q^{(n)} + Q_1^{(n)}$$

$$R^{(n+1)} = T_1^{(n)} + r_1^{(n)} + T_2^{(n)} + \ln\left(1 + \sum\nolimits_2 \sum\nolimits_1^{-1}\right)$$

$$L_{n+1} = L_n^2 2c^{-1/2} \sqrt{\frac{\pi}{2a_0(\beta)}} e^{b_0^{(n)}}.$$

If n is sufficiently large, then $B_{2k}^{(n+1)}$ satisfies the required inequalities for $k=2, \ldots, N$. The constant λ_1 should be chosen close to 1 so that $c\lambda_1>1$.

Let us examine the decrease of $Q^{(n+1)}$. From the recursive equation satisfied by $Q^{(n)}$ it can be seen that

$$Q^{(n+1)} = A^{n-n_0+1}(\beta)(Q^{(n_0)})+\sum_{p=n_0}^{n} A^{n-p}(\beta)(Q_1^{(p)}).$$

Let $\mathscr{H}_N^{\perp}$ be the subspace of $L^2(R^1, e^{-\gamma\beta z^2})$ consisting of all even functions f such that

$$\int f(z)G_{2k}(z;\beta)e^{-\gamma\beta z^2}dz = 0, \quad k = 0, 1, \ldots, N.$$

It is clear that $\mathscr{H}_N^{\perp}$ is invariant under the action of the operator $A(\beta)$ and $\|A\|_{\mathscr{H}_N^{\perp}}= \frac{2}{c^{N+1}}$. Since every $Q_1^{(p)}\in\mathscr{H}_N^{\perp}$, the following relations hold:

$$A^{n-p}(Q_1^{(p)})\in\mathscr{H}_N^{\perp} \quad\text{and}\quad \|A^{n-p}(Q_1^{(p)})\| \leqq (2c^{-N-1})^{n-p}\|Q_1^{(p)}\|.$$

Lemma 4.1. *Let us suppose that t is a function belonging to $\mathscr{H}_N^{\perp}$ and $|t(z)|\leqq 1$ for every z. In addition, the inequalities $0<M_1<\frac{1}{6}\ln c$, $0<M_2<\sqrt{\frac{\ln c}{\text{const}}}$ hold, where an absolute* const *will be given in the course of the proof. Then for sufficiently large N, for any $p>1$ and for $|z|\leqq M_2\sqrt{pN}$ the inequality*

$$|(A^p t)(z)| \leqq e^{-M_1 p(N+1)}$$

is true.

Proof. The following statements are obvious:

1) if $|t(z)|\leqq 1$, then there exists an absolute constant $L_1>1$ such that

$$|A(t)(z)| \leqq L_1, \quad \left|\frac{d}{dz}(At)(z)\right| \leqq L_1$$

2) $\|A^p t\|\leqq(2c^{-N-1})^p\|t\|$.

Here $\|\cdot\|$ denotes the norm in the Hilbert space $L^2(R^1, e^{-\gamma\beta z^2})$.

Now let us assume that for some z_0, $|z_0|\leqq M_2\sqrt{pN}$; the following inequality holds:

$$(A^p t)(z_0) \geqq e^{-M_1 p(N+1)}.$$

The case when $(A^p t)(z_0)$ is negative can be treated analogously. It follows from 1) that

$$\left|\frac{d}{dz}A^p t\right| \leqq L_1^p.$$

Let us construct the function $g(z)$:

$$g(z) = \begin{cases} (A^p t)(z_0)-L_1^p|z-z_0| & \text{for} \quad |z-z_0| \leqq \dfrac{e^{-M_1 p(N+1)}}{L_1^p} \\ 0 \quad \text{elsewhere.} \end{cases}$$

Then

$$0 \leqq g(z) \leqq (A^p t)(z) \quad \text{for} \quad |z-z_0| \leqq \frac{e^{-M_1 p(N+1)}}{L_1^p} = \Gamma$$

hence

$$(2c^{-N-1})^p \geqq \sqrt{\int_{-\infty}^{\infty} [(A^p t)(z)]^2 e^{-\beta\gamma z^2} dz} \geqq$$

$$\geqq \min_{z:\,|z-z_0|<\Gamma} e^{-\beta\gamma z^2} \sqrt{\int_{-\infty}^{\infty} g^2(z)\, dz} \geqq$$

$$\geqq \text{const} \cdot \exp\left\{-\text{const} \cdot \left(M_2 \sqrt{pN}+\Gamma\right)^2\right\} \Gamma \cdot e^{-M_1 p(N+1)}.$$

Let N be so large that $\Gamma \leqq M_2 e^{-\frac{1}{2}M_1 p(N+1)}$. Then

$$\exp\left\{-\text{const} \cdot \left(M_2 \sqrt{pN}+\Gamma\right)^2\right\} \geqq \exp\left\{-\text{const} \cdot M_2^2 pN\right\}$$

and

$$\exp\left\{-\text{const} \cdot \left(M_2 \sqrt{pN}+\Gamma\right)^2\right\} \Gamma e^{-M_1 p(N+1)} \geqq$$

$$\geqq \exp\left\{-\text{const} \cdot M_2^2 pN - p\,\text{const} - 2M_1 p(N+1) + \text{const}\right\}.$$

Here the expressions containing L_1 are included in const.

Hence

$$p(N+1) \ln c + p\,\text{const} \leqq \text{const}\, M_2^2 pN + 2M_1 p(N+1) + \text{const}.$$

For large N the left-hand side is asymptotically equivalent to $pN \ln c$. By the conditions on M_1 and M_2 for large N, the right-hand side is less than the left-hand side. This contradiction proves the lemma.

Lemma 4.1 is a theorem of Tauberian type for Hermite polynomials. It shows that if the Fourier coefficients with respect to the Hermite polynomials are small, then the function is also small on an interval depending on the Fourier coefficients.

Now we can estimate $Q^{(n+1)}$ by means of Lemma 4.1. Let n_1 be equal to $[n\omega]$ for some ω, $0<\omega<1$. If $p \leqq n_1$, then $n-p \geqq (1-\omega)n$, $M_2\sqrt{(n-p)N} \geqq M_2\sqrt{(1-\omega)N} \cdot \sqrt{n}$. Let us choose N so large that $M_2\sqrt{(1-\omega)N} > D$. Under this condition we can estimate $A^{n-p}(\beta)Q_1^{(p)}$ by the lemma with the result

$$\lambda_1^p (2c^{-2})^p \exp\{-M_1(n-p)(N+1)\} \leqq \lambda_1^p (2c^{-2})^p \exp\{-nM_1(1-\omega)(N+1)\}.$$

For every ω we can choose N in such a way that

$$\exp\{-M_1\omega(N+1)\} \leqq \lambda_1^2(2c^{-2}).$$

An analogous inequality is true for $p=n_0$. As a result we get

$$|A^{n-n_0+1}(\beta)Q^{(n_0)}| + \sum_{p=n_0}^{n_1} |A^{n-p}(\beta)Q_1^{(p)}| \leqq$$

$$\leqq \text{const} \cdot \lambda_1^n (\lambda_1 2c^{-2})^n.$$

For p, $n_1 < p \leqq n$ we have

$$|A^{n-p} Q_1^{(p)}| \leqq L_1^{n-p} |Q_1^{(p)}| \leqq \text{const} \cdot L_1^{n(1-\omega)} n^N \lambda_1^{3p/2} (2c^{-2})^p.$$

By summation we get

$$\sum_{p=n_1+1}^{n} |A^{n-p} Q_1^{(p)}| \leqq \text{const} \cdot L_1^{n(1-\omega)} n^N \lambda_1^{3n_1/2} (2c^{-2})^{n_1}.$$

Let us choose ω sufficiently close to 1 so that

$$L_1^{1-\omega} \lambda_1^{3\omega/2} (2c^{-2})^{\omega} < \lambda_1 (2c^{-2}).$$

Then for n large enough the last expression will be less than $\frac{1}{2} \lambda_1^{n+1} (2c^{-2})^{n+1}$. Adding this to the estimation obtained for $A^{n-n_0+1} Q^{(n_0)} + \sum_{p=n_0}^{n} A^{n-p} Q_1^{(p)}$ we get the required estimation for $Q^{(n+1)}$.

Also, the relation $\frac{d}{dz} Q^{(n+1)}(z) \leqq \lambda_1^{3n/2} (2c^{-2})^n$ holds if $z \in D_{n+1}$. The proof is analogous to the proof of the previous estimation with the only difference that the term $\frac{d}{dz} Q_1^{(n)}(z)$ should be considered separately. But

$$\left| \frac{d}{dz} Q_1^{(n)}(z) \right| = \left| \frac{d}{dz} A(\beta) R^{(n)} - \sum_{k=0}^{N} b_{2k}^{(n)} G_{2k}(z; \beta) \right| \leqq$$
$$\leqq \text{const} \cdot n^{2N} (2c^{-2}) \lambda_1^{8n/2}.$$

Now we are going to deal with the estimation of the errors. Integrating (4.12) over the interval $|u| < \frac{D}{6\sqrt{c}} \sqrt{n}$ with respect to the weight function $e^{-2a_0(\beta) u^2}$ we get that

$$|T_1^{(n)}| \leqq L_1 \cdot \text{const} \cdot B^2 n^N (2c^{-2})^{2n} = L_1 \cdot \text{const} \cdot B^2 n^N (2c^{-2})^n (2c^{-2})^n.$$

Let us choose λ_1 in such a way that $2c^{-2} < \lambda_1^{3/2} < 1$. Then for sufficiently large n

$$|T_1^{(n)}| < \frac{1}{5} \lambda_1^{3/2} (2c^{-2})^{n+1}. \tag{4.14}$$

The error $r_1^{(n)}$ arises from the substitution of the Riemannian sum by an integral. In virtue of the choice of the interpolation formula, this error can be made very small, when products of an exponential function and a polynomial are summed (order of $\Delta_n^2 = c^n 2^{-2n}$). When we substitute the sum by an integral in the expression containing $R^{(n)}$, $T^{(n)}$, the error can be estimated by the product of the maximum of these quantities and the step length $\Delta_n = c^{n/2} 2^{-n}$ since they are step functions. The expression containing $Q^{(n)}$ can similarly be estimated applying the upper bound obtained for

$\frac{dQ^{(n)}}{dz}$. It follows from these estimations that

$$|r_1^{(n)}| \leqq \text{const}\cdot[\lambda_1^n(2c^{-2})^n\Delta_n+\Delta_n^2].$$

Let us turn to the estimation of $r_2^{(n)}$. When the question is to extend the integral of expressions containing a product of a polynomial of degree not higher than $2N$ and an exponential function from the interval $|u|\leqq\frac{D}{6\sqrt{c}}\sqrt{n}$ to the whole real line, then the error does not exceed $C_1n^Ne^{-C_2D^2n}$, where C_1 depends on N, but C_2 is independent of it. Let us choose D so large that $e^{-D^2C_2}<\lambda_1^2(2c^{-2})$. Then for every N there exists a natural number $n(N)$ such that the above treated part of the error $r_2^{(n)}$ does not exceed $\lambda_1^{2n}(2c^{-2})^n$ for $n\geqq n(N)$. Somewhat finer consideration is required to treat the error arising from the integration of $Q^{(n)}$. Let us take into account again that $Q^{(n)}=A^{n-n_0}Q^{(n_0)}\pm\sum_{p=0}^{n-1}A^{n-p}Q_1^{(p)}$. We have already shown (see 4.13) that

$$|Q_1^{(p)}|\leqq\text{const}\cdot p^{2N}\lambda_1^{3p/2}(2c^{-2})^p;$$

therefore

$$|A^{n-p}Q_1^{(p)}|\leqq L_1^{n-p}\cdot\text{const}.\, p^N\lambda_1^{3p/2}(2c^{-2})^p\leqq\text{const}\cdot L_1^n n^N\lambda_1^{3p/2}(2c^{-2})^p.$$

Thus the part arising from the term $A^{n-p}Q_1^{(p)}$ of the error $r_2^{(n)}$ does not exceed $\text{const}\cdot L_1^n e^{-D^2\,\text{const}\cdot n}n^{2N}\lambda_1^{3p/2}(2c^{-2})^p$. Summing for p we get again that if D is sufficiently large (larger than some absolute constant), then for sufficiently large n the error $r_2^{(n)}$ will be less than $\frac{1}{5}\lambda_1^{3n/2}(2c^{-2})^n$.

The last step is to estimate $\ln(1+\sum_2\sum_1^{-1})$. The smallness of this quantity follows from the external estimation $\text{Ind}_3^{(n)}$ and the rapid decrease of the Gaussian density. Applying $\text{Ind}_1^{(n)}$ for $\frac{z}{\sqrt{c}}\pm u\in D_n$ and $\text{Ind}_3^{(n)}$ for $\frac{z}{\sqrt{c}}\pm u\notin D_n$ we obtain the inequality

$$\sum\nolimits_2=\exp(\beta z^2)\sum_{|u|>\frac{D}{6\sqrt{c}}\sqrt{n}}Z_n\left(\frac{z}{\sqrt{c}}+u;\,\beta\right)Z_n\left(\frac{z}{\sqrt{c}}-u;\,\beta\right)\leqq$$

$$\leqq L_n^2(\beta)c^{n/2}2^{-n}\exp\{-a_0(\beta)z^2+\text{const}_1\cdot(2c^{-2})^nD_n^2-\text{const}_2\cdot D_n^2\}.$$

On the other hand

$$\sum\nolimits_1\geqq\text{const}\cdot L_n^2(\beta)\Delta_n\exp\{-a_0(\beta)z^2+\text{const}_3\cdot(2c^{-2})^nD_n^2\}.$$

Therefore $\sum_2\sum_1^{-1}\leqq\text{const}\cdot\exp\{-\text{const}\cdot D_n^2+\text{const}_4\cdot(2c^{-2})^nD_n^2\}$. Thus choosing D and n_0 sufficiently large, we can achieve that the last error be less than $\frac{1}{5}\lambda_1^{3(n+1)/2}(2c^{-2})^{n+1}$ for $n>n_0$. Summing all these estimations we get that $R^{(n+1)}$ satisfies the required estimation in $\text{Ind}_1^{(n+1)}$.

Thus for any $\beta^{(n+1)} \subset \beta^{(n)}$ the condition $\mathrm{Ind}_1^{(n+1)}$ will be fulfilled. Let us observe that $B_2^{(n+1)}(\beta_+^{(n)}) > \lambda_1^{n+1} c^{-n-1}$ and $B_2^{(n+1)}(\beta_-^{(n)}) < -\lambda_1^{n+1} c^{-n-1}$. Therefore we can choose the numbers $\beta_+^{(n+1)}$ and $\beta_-^{(n+1)}$ satisfying $\mathrm{Ind}_2^{(n+1)}$ in such a way that $[\beta_-^{(n+1)}, \beta_+^{(n+1)}] \subset \subset [\beta_-^{(n)}, \beta_+^{(n)}]$. For let us consider $B_2^{(n+1)}$. In the main order $B_2^{(n+1)} = B_2^{(n)}$. The addition of $b_2^{(n)}$ changes $B_4^{(n)}$ by only a relatively small amount at least at the ends of the interval $\beta^{(n)}$. It follows from $\mathrm{Ind}_4^{(n)}$ that $b_2^{(n)}$ and $B_2^{(n+1)}(\beta)$ are continuous functions of β if $\beta \in \beta^{(n+1)}$. Let us choose $\beta^{(n+1)}$ in such a way that $\mathrm{Ind}_2^{(n+1)}$ be satisfied by $B_2^{(n+1)}(\beta)$. In spite of its simplicity there is a deeper meaning to this consideration; firstly it describes the process of finding β_{cr}; secondly, from the point of view of the geometrical picture given above it is a mean of determining the value β for which the corresponding probability distribution would lie on the stable manifold of the given Gaussian solution of the basic integral equation.

The last step is to check $\mathrm{Ind}_3^{(n+1)}$. For $z > D\sqrt{n+1}$ we have (the case $z < -D\sqrt{n+1}$ can be treated analogously):

$$Z_{n+1}(z;\beta) \leqq L_n^2 2^{-n-1} c^{(n+1)/2} e^{-a_0(\beta)z^2} \cdot$$

$$\cdot \Bigg[\sum_{\substack{\left|\frac{z}{\sqrt{c}}+u\right| \leqq \frac{D}{2}\sqrt{n} \\ \left|\frac{z}{\sqrt{c}}-u\right| > \frac{D}{2}\sqrt{n}}} \Delta_n e^{-2a_0(\beta)u^2} e^{-\bar{B}_4^{(n)}(2c^{-2})^n (z/\sqrt{c}+u)^4 - \bar{B}_4^{(n)}(2c^{-2})^n (z/\sqrt{c}-u)^4} +$$

$$+2 \sum_{\substack{\left|\frac{z}{\sqrt{c}}+u\right| > \frac{D}{2\sqrt{c}}\sqrt{n} \\ \left|\frac{z}{\sqrt{c}}-u\right| > \frac{D}{2\sqrt{c}}\sqrt{n}}} e^{-2a_0(\beta)u^2 - \bar{B}_4^{(n)}(2c^{-2})^n ((z/\sqrt{c}+u)^4 + (z/\sqrt{c}-u)^4) + B_4^{(n)} D_n^2 (2c^{-2})^n} \Bigg] \leqq$$

$$\leqq L_n^2 \Delta_{n+1} 2c^{-1/2} e^{-a_0(\beta)z^2 - \bar{B}_4^{(n)}(2c^{-2})^{n+1} z^4} (I_1 + 2I_2).$$

The sum I_1 can, just as above, be estimated by reduction to the Gaussian integral, which gives $I_1 \leqq \sqrt{\dfrac{\pi}{2a_0(\beta)}}(1+\delta_1^n)$. The error $|\delta_1^{(n)}|$ does not exceed const $\cdot\, 2^{-n}$. From the inequality $z \leqq D\sqrt{n+1}$ we obtain for I_2 that $|u| \geqq \dfrac{D}{2\sqrt{c}}\sqrt{n}$ and so $I_2 \leqq \leqq \exp\{-\text{const} \cdot D_n^2\}$. Finally we get

$$2I_2 + I_1 \leqq (1 + \text{const} \cdot 2^{-n}) \sqrt{\frac{\pi}{2a_0(\beta)}}.$$

Therefore applying the recursion formula for L_{n+1} we obtain that

$$Z_{n+1}(z,\beta) \leqq L_{n+1}(\beta)(2c^{-1/2})^{n+1} \exp\{-a_0(\beta)z^2 - \bar{B}_4^{(n)}(2c^{-2})^{n+1} z^4 (1-\alpha_{n+1})\},$$

where $\alpha_{n+1} = \dfrac{\text{const} \cdot \Delta_n^2 - b_0^{(n)}}{\bar{B}_4^{(n)}(2c^{-2})^{n+1} z^4}$.

By means of the estimation for $b_0^{(n)}$ we get that

$$\alpha_{n+1} \leqq \text{const} \cdot \frac{n^N \cdot \lambda_1^{3n/2}}{z^4 \bar{B}_4^{(n)}} \leqq \text{const} \cdot \frac{n^N \lambda_1^{3n/2}}{D^4 \bar{B}_4^{(n)}} \leqq n^N \lambda_1^{3n/2}$$

if D is sufficiently large. Thus $\text{Ind}_3^{(n+1)}$ holds with $\bar{B}_4^{(n+1)} = \bar{B}_4^{(n)}(1 - n^N \lambda_1^{3n/2})$. The lemma is proved.

Remark. The basic parameters used in the proof were the constants λ_1, D, N, n_0. The parameter λ_1 was restricted twice in the course of the proof. After the choice of λ_1, D was chosen. The constraints to be made for D were indicated by the speed of the decrease of the Gaussian density at infinity. Then we choose N to ensure the required smallness of $Q^{(n)}$. Finally a sufficiently large n_0 was chosen.

Let us now finish the proof of the theorem. The induction hypotheses $\text{Ind}^{(n)}$ are fulfilled for $\beta_{\text{cr}} \in \bigcap_n \beta^{(n)}$ when $n \geqq n_0$. We show that the discrete probability distribution with probabilities of the form

$$\frac{Z_n(z;\, \beta_{\text{cr}})}{\Theta(\beta_{\text{cr}})} = g_n(z;\, \beta_{\text{cr}})(c^{1/2}2^{-1})^n$$

converges weakly to the probability distribution with the Gaussian density $\sqrt{\frac{a_0(\beta_{\text{cr}})}{\pi}}\, e^{-a_0(\beta_{\text{cr}})z^2}$. Indeed for $z \in D^{(n)}$

$$Z_n(z;\, \beta_{\text{cr}}) = L_n(\beta_{\text{cr}}) \Delta_n e^{-a_0(\beta_{\text{cr}})z^2} e^{-\sum_{k=1}^N B_{2k}^{(n)}(\beta_{\text{cr}}) G_{2k}(z;\beta_{\text{cr}}) - Q^{(n)} - R^{(n)}}.$$

In virtue of $\text{Ind}_1^{(n)}$ the last factor tends to 1 uniformly with respect to $z \in D^{(n)}$ for $n \to \infty$. It follows from the external estimation that for $z \notin D^{(n)}$

$$Z_n(z;\, \beta_{\text{cr}}) \leqq 2L_n(\beta_{\text{cr}}) \Delta_n e^{-a_0(\beta_{\text{cr}})z^2 - (2c^{-2})^n \bar{B}_4^{(n)} z^4}.$$

Here we used the inequality $0 < \bar{B}_4^{(n)}$. Hence, taking into account the normalization, we get the weak convergence of $g_n(z;\, \beta_{\text{cr}})$ to the Gaussian limit.

Now we shall discuss the initial Hamiltonian. For $n = n_0$ we have a volume of 2^{n_0} points. Let us write the interaction potential as

$$\mathscr{H}_{n_0}(\varphi(V^{(n_0)})) = C_1 \Phi^2(\varphi(V^{(n_0)})) - C_2 \Phi^4(\varphi(V^{(n_0)})),$$
$$\Phi(\varphi) = \sum_{x \in V^{(n_0)}} \varphi(x).$$

The positive constants C_1 and C_2 will be determined below. We have

$$Z_{n_0}(t;\, \beta) = e^{-\beta(C_1 t^2 + C_2 t^4)} \begin{pmatrix} 2^{n_0} \\ 2^{n_0 - 1} - \frac{t}{2} \end{pmatrix}.$$

For any t of order $2^{n_0} c^{-\frac{n_0}{2}}$, by setting $\xi = t2^{-n_0}$ we have the following expansion of the binomial coefficient

$$\binom{2^{n_0}}{2^{n_0-1}-\frac{t}{2}} = 2^{n_0}\, e^{2^{n_0}(a_0-a_2\xi^2-a_4\xi^4-\ldots-a_{2N}\xi^{2N}+\mathcal{O}(\xi^{2N+2}))}.$$

Using the variable $z=\xi c^{n/2}$ we obtain

$$Z_{n_0}(z;\beta) = L_n' \exp\{\beta(C_1 c^{-n_0}2^{2n_0}z^2+C_2 c^{-2n_0}2^{4n_0}z^4)-a_2 2^{n_0}c^{-n_0}z^2-$$
$$-a_4 2^{n_0}c^{-2n_0}z^4-a_6 2^{n_0}c^{-3n_0}z^6-\ldots-a_{2N}2^{n_0}c^{-Nn_0}z^{2N}+2^{n_0}\mathcal{O}(\xi^2)\}.$$

It is easy to see that choosing the constant C_1 conveniently the exponent can be written as

$$\text{const}\cdot -a_0(\beta)z^2+(C_3\beta-a_2' 2^{n_0}c^{-n_0})G_2(z;\beta)+$$
$$+(\beta C_2 c^{-2n_0}2^{4n_0}-a_4' 2^{n_0}c^{-2n_0})G_4(z;\beta)+\ldots$$

where dots mean the Hermite polynomials of higher degree and the error.

For any given $\beta_{cr}^{(0)}$ let us choose C_3 depending on C_1 in such a way that $C_3\beta_{cr}^{(0)}-a_2 2^{n_0}c^{-n_0}=0$. Then on account of the choice of n_0, $\beta_{cr}^{(0)}$ can be made arbitrarily close to β_{cr}. The quantity C_2 should be taken so that the coefficient of G_4 be negative and less than $-\text{const}\cdot(2c^{-2})^{n_0}$. Clearly, if $\beta=\beta_{cr}^{(0)}$, then $\mathcal{H}_{n_0}$ satisfies $\text{Ind}_1^{(n_0)}$ (with $B_2^{(n_0)}=0$, $B_6^{(n_0)}=0, \ldots, Q^{(n_0)}(z)=0$), $\text{Ind}_3^{(n_0)}$ and $\text{Ind}_4^{(n_0)}$. It can be checked that $\mathcal{H}_{n_0}$ also fulfils them for $\beta\in[\beta_{cr}^{(0)}-\alpha c^{-n}, \beta_{cr}^{(0)}+\alpha c^{-n}]$, if α is sufficiently small. Then $\beta_2^{(n_0)}\sim\text{const}\cdot(\beta-\beta_{cr}^{(0)})$ in $\text{Ind}_1^{(n_0)}$, so $\text{Ind}_2^{(n_0)}$ can be satisfied. Obviously the Hamiltonians sufficiently close to $\mathcal{H}_{n_0}$ also fulfil $\text{Ind}^{(n_0)}$.

The remaining cases are $\beta<\beta_{cr}$ and $\beta>\beta_{cr}$. We shall not carry out the complete proofs, but restrict ourselves to an overall description of the necessary considerations. First let us assume that β is less than β_{cr} but sufficiently close to it in order to ensure that, for some $\bar{n}=\bar{n}(\beta)$, $\beta\in\beta^{(\bar{n})}$. We can use the induction hypotheses $\text{Ind}^{(n)}$ up to $\bar{n}$. The assertion that the function $B_2^{(n)}(\beta)$ is differentiable with respect to β can be shown by a similar but slightly more precise consideration than that applied in the proof of Lemma 4.1. In addition the derivative remains bounded from above by a negative constant, therefore $B_2^{(\bar{n})}(\beta)\sim\text{const}\cdot(\beta-\beta_{cr})$, for $\beta\to\beta_{cr}$. If $n>\bar{n}(\beta)$, then we have to give an other representation to $Z_n(t;\beta)$*:

$$Z_n(t;\beta) = L_n(\beta)2^{-n/2}e^{-2^n\left(a_0(\beta)\left(\frac{c}{2}\right)^n(2^{-n}t)^2+B_2^{(n)}(\beta)(2^{-n}t)^2+P^{(n)}(t;\beta)\right)}.$$

It can be shown that $P^{(n)}$ tends to 0 uniformly in t for $|t|\leq\dfrac{\text{const}\cdot\sqrt{n}\,2^{n/2}}{\sqrt{\beta_{cr}-\beta}}$. As a

* The use of the coordinate t instead of z shows that we returned to the values of the complete spin again.

result we get that $\frac{Z_n(z2^{-n/2};\,\beta)}{\Theta_n(\beta)}\,2^{n/2}$ converges weakly to the Gaussian density with dispersion of order $\frac{\text{const}}{\beta_{\text{cr}}-\beta}$. This means that the critical index γ is equal to 1.

Since the accurate considerations concerning $\beta>\beta_{\text{cr}}$ are more cumbersome, we explain only the idea. For this case the inductive hypotheses also can be applied up to a natural number $\bar{\bar{n}}=\bar{\bar{n}}(\beta)$. (During the recursion the coefficient of z^2 increases. The recursion can be executed up to the $\bar{\bar{n}}$th step, when the coefficient of z^2 becomes positive first.)

For $n=\bar{\bar{n}}$ we write $z_{\bar{\bar{n}}}(t;\,\beta)$ in the form

$$Z_{\bar{\bar{n}}}(t;\,\beta)=L_{\bar{\bar{n}}}(\beta)\exp\Big\{-2^{\bar{\bar{n}}}\big(b_1^{(\bar{\bar{n}})}(\beta)+b_2^{(\bar{\bar{n}})}(\beta)\xi^2+b_3^{(\bar{\bar{n}})}\xi^4+ \\ +a_0(\beta)\,c^{\bar{\bar{n}}}\,2^{-\bar{\bar{n}}}\xi^2\big)+P_1^{(\bar{\bar{n}})}(\xi;\,\beta)\big)\Big\},\quad \xi=t2^{-\bar{\bar{n}}},$$

where $P_1^{(n)}$ is to be regarded as an error term. The quantity $b_2^{(\bar{\bar{n}})}(\beta)$ is equivalent to $\text{const}\cdot(\beta_{\text{cr}}-\beta)$, and n is such that $|b_2^{(\bar{\bar{n}})}(\beta)|>c^{\bar{\bar{n}}}2^{-\bar{\bar{n}}}$. This means that

$$b_1^{(\bar{\bar{n}})}(\beta)+\big(b_2^{(\bar{\bar{n}})}(\beta)+a_0(\beta)\,c^{\bar{\bar{n}}}\,2^{-\bar{\bar{n}}}\big)\xi^2+b_3^{(\bar{\bar{n}})}\xi^4,$$

has the form pictured in Figure 4.2, and thus it has two minima at the points $\pm\text{const}\cdot\sqrt{-b_2^{(\bar{\bar{n}})}(\beta)+a_0(\beta)c^{\bar{\bar{n}}}2^{-\bar{\bar{n}}}}$. Precise considerations show that for $n>\bar{\bar{n}}$ the probability distribution f_n asymptotically (for $n\to\infty$) behaves as a "half-sum" of two Gaussian distributions concentrated around the points $\xi=\pm\text{const}\cdot\sqrt{\beta-\beta_{\text{cr}}}$. (Since we need a good estimate of Z_n in a region containing its maxima – which are not contained in D_n–large deviation results must be applied in the proof.) This means that the critical index ω is equal to $\frac{1}{2}$.

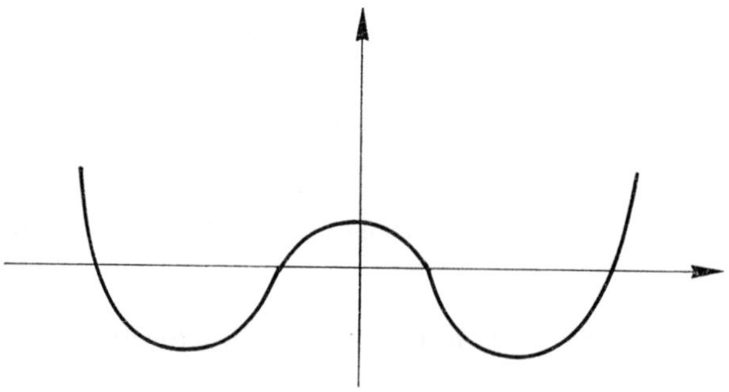

FIGURE 4.2

4. The domain $c<\sqrt{2}$

For $c=\sqrt{2}$ the analysis given in Section 3 can be easily improved to include this case too. But for $c<\sqrt{2}$ the eigenvector G_2 has the eigenvalue $2c^{-2}>1$ and therefore there is no reason to expect that the Gaussian solution be thermodynamically stable.

It turns out that for c sufficiently close to $\sqrt{2}$, Equation (4.6) has a non-Gaussian solution, which is thermodynamically stable. The complete proof of this statement is very cumbersome, and not given here (see [48], [119]). We shall only describe the main ideas based on the theory of bifurcations and the theory of the invariant manifolds of the fixed-points of diffeomorphisms.

By the substitution $g(z;\beta)=h(z;\beta)\exp\left\{-\frac{\beta c}{2-c}z^2\right\}$ Equation (4.6) becomes

$$(4.6') \qquad h(z;\beta)=\int_{-\infty}^{\infty} h\left(\frac{z}{\sqrt{c}}+u;\beta\right)h\left(\frac{z}{\sqrt{c}}-u;\beta\right)e^{-\frac{2\beta c}{2-c}u^2}\,du.$$

As we have seen it is sufficient to solve (4.6′) for an arbitrary value of β. Particularly, one can choose $\beta_0=(2-c)(2c)^{-1}$. Then we get the equation for $h(z)=h(z;\beta_0)$

$$(4.6'') \qquad h(z)=\int_{-\infty}^{\infty} h\left(\frac{z}{\sqrt{c}}+u\right)h\left(\frac{z}{\sqrt{c}}-u\right)e^{-u^2}\,du,$$

which can be considered as an equation for the fixed point of the quadratic transformation $T_c(h)$ on the right-hand side. Equation (4.6″) has a solution $h(z)=\sqrt{\pi}$ independent of c. For $c=\sqrt{2}$ in the spectrum of the differential $\partial T_c(h)=Ah$ there appears the value 1, the corresponding eigenvector is the Hermite polynomial G_4. If the projection to G_4 of the result of the application of the second differential to G_4 is different from 0, then on the basis of the general theory of bifurcations in finite-dimensional spaces we can expect a new fixed point to appear. However, neither the finite-dimensional theory, nor its generalizations to Banach spaces can directly be applied, principally because the set of values z is not compact.

This can in particular be deduced from the fact that for $c>\sqrt{2}$ there are no non-Gaussian solutions of (4.6″) close to the Gaussian one. Nevertheless we can construct the non-Gaussian solutions by modifying the methods of the theory of the invariant manifolds of fixed points of nonlinear transformations.

In order to show how it can be made, let us notice that the perturbation theory according to the parameter $\varepsilon=\sqrt{2}-c$ permits expressions for the fixed point to be derived up to an arbitrary order in ε. In fact it is sufficient to go up to ε^2. So we get an expression as follows

$$g_\varepsilon(z)=e^{-\frac{1}{2}z^2-\varepsilon a G_4(z)-\varepsilon^2 P_8(z)}, \quad g_\varepsilon(z)=g(z;\beta_0),\quad c=\sqrt{2}-\varepsilon,$$

where $P_8(z)$ is a polynomial of degree eight. It also follows from the perturbation theory that in the neighbourhood of the unknown solution the spectrum of the differential of the transformation has an eigenvalue less than 1 (more precisely of order $1-\text{const}\cdot\varepsilon$) and the rest of the spectrum lies in a circle of radius less than 1 and independent of ε. This means that the unknown solution is an unstable but hyperbolic fixed point. We can find such a fixed point in the following way: we choose some initial curve going in a direction close to the direction of the unstable eigenvector of the unknown fixed point. (The eigenvector can also be found on the basis of perturbation theory.) Let us assume that this curve is sufficiently close to the unknown fixed point. In our case it is easy to construct such a curve. If we choose the curve successfully, then it intersects the stable manifold of the unknown point* exactly once. From this picture a method follows for constructing the fixed point: after having chosen an initial curve let us apply the transformation T_c to it and keep only the part lying in a given, sufficiently small neighbourhood of g_ε. Then there exists only one point on the initial curve remaining in this neighbourhood for all iterations. This point is the intersection of the initial curve with the stable manifold, and it converges to the unknown fixed point under the action of the powers of T_c.

The method described above does indeed work. It leads to an existence proof for the solution $h(z)$ which decreases at infinity as $\exp\{-\text{const}\cdot\varepsilon z^\delta\}$, $\delta=\frac{2}{\log_2 c}\sim 4$, for small ε. Thus the density of the limit distribution $h(z;\beta)$ decreases faster than the Gaussian one for $|z|\to\infty$.

In the non-Gaussian case the critical indices depend on c. It turns out that if $d=2(1-\log_2 c)^{-1}$ is the dimension of the system, then the indices are $\eta=0$, $\gamma=\log_{\lambda_1}\frac{2}{c}$, $\omega=\frac{1}{2}\log_{\lambda_1} c$, where $\lambda_1>1$ is the unstable eigenvalue corresponding to the solution obtained. The first terms of the expansion of λ_1 with respect to the parameter ε were found in [48]. In an interesting paper of Collet and Eckman [54] it was shown that λ_1 is an infinitely differentiable function of ε.

In connection with the results presented above two problems arise quite naturally:

1. To prove that the branch of solutions g_ε constructed by means of perturbation theory can be extended to the end point, i.e. to $\varepsilon=\sqrt{2}$; numerical computations made by P. M. Bleher (see [48]) convince us that this is really so. It is quite possible that the numerical results obtained by computer can be used for the rigorous mathematical solution of the problem.

* Remember that the stable manifold of a fixed point consists of the points that converge to the fixed point when the transformation is iterated.

2. The investigations of the analytical nature of λ_1 in the neighbourhood of $\varepsilon=0$ are not yet finished. In particular, it cannot be excluded that the formal series of λ_1 with respect to ε is Borel summable. The clarification of this question would be important for the investigation of the nature of ε-expansions in the theory of phase transitions.

5. Scaling probability distributions

As we have pointed out in Section 1 the ferromagnetic critical point β_{cr} can be characterized in the following way: the correlations $\langle\varphi(x), \varphi(y)\rangle$ decrease for $V\to\infty$ so slowly that $\langle[\Phi(\varphi(V))]^2\rangle \sim \text{const}\cdot V^\alpha$ for $\beta=\beta_{cr}$ where α is a parameter, $1<\alpha<2$. The expectations are taken with respect to the limit Gibbs distribution P_1 corresponding to the Hamiltonian $\mathcal{H}$. Let us assume that P_1 is translation-invariant. Let us decompose the lattice Z^d into cubes $\Delta_k(x)$ containing k^d points

$$\Delta_k(x) = \{y\in Z^d:\ kx_i \leqq y_i < k(x_i+1),\ i = 1, \dots, d\};$$

the family of cubes $\Delta_k(x)$ forms a lattice again. Let $\varphi_k(x)$ be equal to $\frac{1}{k^{d\alpha/2}}\sum_{y\in\Delta_k(x)}\varphi(y)$ and let P_k be the probability distribution induced by the random variables $\varphi_k(x)$. It is clear that P_k is also translation-invariant. The following assumption is considered to be reasonable in the theory of critical indices: the probability distributions P_k converge weakly to a limit for $k\to\infty$. The limit distribution Q weakly depends on the initial Hamiltonian and is universal in some sense.

According to the usual method of probability theory we first determine what kind of distributions can be limit distributions at all. Let us denote by $\mathfrak{M}$ the vector space of real-valued sequences $\varphi=\{\varphi(x), x\in Z^d\}$. In an obvious way we can define the action of the group of the space translations $\{T^t, t\in Z^d\}$ on $\mathfrak{M}$: $(T^t\varphi)(x)=\varphi(x+t)$. Let k be a natural number. Let us introduce the linear endomorphism $\mathfrak{A}_k(\alpha)=\mathfrak{A}_k$ of the space $\mathfrak{M}$ as follows:

$$(\mathfrak{A}_k\varphi)(x) = \frac{1}{k^{d\alpha/2}}\sum_{y\in\Delta_k(x)}\varphi(y).$$

The endomorphisms $\mathfrak{A}_k$ form a multiplicative semigroup: $\mathfrak{A}_{k_1}\cdot\mathfrak{A}_{k_2}=\mathfrak{A}_{k_1\cdot k_2}$. The semigroup $\mathfrak{A}=\{\mathfrak{A}_k\}$ is an appropriate tool to study the behaviour of normalized random variables.

Definition 4.3. *The semigroup $\mathfrak{A}$ is called the renormalization group.*

It is clear that $T^t\mathfrak{A}_k=\mathfrak{A}_k T^{tk}$. Let us denote by $\mathfrak{A}^{(0)}$ the semigroup generated by all transformations T^t and $\mathfrak{A}_k$.

Lemma 4.2. *The semigroup $\mathfrak{A}^{(0)}$ is isomorphic to the semigroup of the linear transformations of the space R^d given by the formula: $\vec{x} \to \frac{\vec{x}+\vec{t}}{k}$ where $\vec{t} \in Z^d$ and k is a natural number.*

Proof. Let the translation $\vec{x} \to \vec{x}+\vec{t}$ correspond to T^t and the contraction $\vec{x} \to \frac{\vec{x}}{k}$ correspond to $\mathfrak{A}_k$. It is easy to see that the translations and the contractions fulfil the same commutation laws as T_t, $\mathfrak{A}_k$. Therefore the semigroup generated by them is isomorphic to $\mathfrak{A}^{(0)}$. The lemma is proved.

Let us denote by $(\mathfrak{A}^{(0)})^*$ the adjoint semigroup acting on the space of all probability distributions on $\mathfrak{M}$, i.e. for every $S \in \mathfrak{A}^{(0)}$ and for every probability distribution P on $\mathfrak{M}$. The lemma is proved.

$$(S^*P)(C) = P(S^{-1}C), \quad C \subset \mathfrak{M}.$$

Definition 4.4. *The probability distribution Q on $\mathfrak{M}$ is said to be scaling iff it is a fixed point of the semigroup $(\mathfrak{A}^{(0)})^*$.*

It is clear that a scaling distribution Q is translation-invariant, and that the limit distributions mentioned at the beginning of this section are scaling. Now our aim is to describe a certain class of the scaling distributions.

This problem, in its full generality, is very complicated since the class of all scaling distributions is rather wide. From the point of view of the theory developed in this chapter the importance of a given scaling distribution depends on the class of Hamiltonians for which this distribution appears as a limit for $\beta=\beta_{cr}$.

Now let us turn to the continuous analogues of the notions introduced above. It is convenient to apply the methods of the theory of generalized stochastic processes. Let a space $\mathscr{F}$ of basic functions defined on R^d be given. Let us assume that $\mathscr{F}$ is invariant under the action of the groups of translations and similarities, i.e.

$$(T^t f)(x) = f(x+t) \in \mathscr{F}, \quad (\mathfrak{A}_\lambda f)(x) = f(\lambda x) = f(\lambda x_1, \ldots, \lambda x_d) \in \mathscr{F}$$

for every $\lambda > 0$ and every $f \in \mathscr{F}$. The group $\mathfrak{A}^{(0)}$, generated by the groups of translations $\{\vec{x} \to \vec{x}+\vec{t}\}$ and similarities $\{\vec{x} \to \lambda\vec{x}\}$, is isomorphic to the group of all linear transformations of R^d of the form $\vec{x} \to \lambda(\vec{x}+\vec{t})$.

Let us consider the action of the group $\mathfrak{A}^{(0)}$ to the space of linear continuous functionals defined on $\mathscr{F}$

$$(\mathfrak{A}\varphi, f) = \lambda^a(\varphi, \mathfrak{A}_f)$$

where $(\mathfrak{A}f)(x) = f(\lambda(x+t))$ corresponds to $\mathfrak{A}\vec{x} = \lambda(\vec{x}+\vec{t})$. The parameter a is analogous to the parameter α of the discrete case. Let us denote by $(\mathfrak{A}^{(0)})^*$ the adjoint group acting on the space of probability distributions on $\mathscr{F}'$, i.e. on the space of generalized stochastic processes.

Definition 4.5. *The probability distribution P on $\mathscr{F}'$ is said to be scaling iff it is invariant with respect to $(\mathfrak{A}^{(0)})^*$, i.e. $(\mathfrak{A}^* P)(C)=P(\mathfrak{A}^{-1}C)$ for every $C\subset\mathscr{F}'$.*

Now we explain the connection between the two definitions. Let $\chi_{\Delta(x)}(y)$ be the indicator function of the set $x_i \leqq y_i < x_i+1$, $i=1, \dots, d$. Let us suppose that there exists a sequence of functions $f_n \in \mathscr{F}$ tending to $\chi_{\Delta(x)}$ in some sense so that the corresponding sequence of random variables (φ, f_n) tends weakly to a certain random variable denoted by $\psi(x)$.

Under natural conditions on the type of this convergence the probability distribution induced by the random variables $\psi(x)$ is scaling with the parameter $\alpha=-2a/d$. In the continuous case the scaling property means that the random variables $\int f(y)\varphi(y)\,dy$ and $\lambda^a \int \varphi(y) f(\lambda y)\,dy$ are identically distributed. Put $\lambda=k^{-1}$ and $f=\chi_{\Delta_k(x)}$ if possible. Then we get that $\int_{\Delta_k(0)} \varphi(y)\,dy$ and $k^{-a}\int_{\Delta_1(0)}\varphi(y)\,dy$ are identically distributed, i.e. we have a discrete-parameter scaling distribution with the parameter $a=-\frac{d\alpha}{2}$. By this observation a lot of formulae, derived later, will be comprehensible.

6. Gaussian scaling distributions

The Gaussian scaling distributions can fairly easily be described. Let us begin with the one-dimensional lattice case $d=1$.

Lemma 4.3. *A stationary Gaussian distribution is scaling if and only if its spectral density $\varrho(\lambda)$ is of the form*

$$\varrho(\lambda) = C\,|e^{2\pi i\lambda}-1|^2 \sum_{m=-\infty}^{\infty} \frac{1}{|\lambda+m|^{\alpha+1}}$$

where $C>0$ is an arbitrary constant.

Proof. First let us verify that a stationary Gaussian distribution with the spectral density as above is scaling. For this purpose we need to show that

$$\left\langle \frac{1}{k^{\alpha/2}} \sum_{s=kx}^{k(x+1)-1} \varphi(s), \frac{1}{k^{\alpha/2}} \sum_{s=0}^{k-1} \varphi(s) \right\rangle = \langle \varphi(x), \varphi(0)\rangle.$$

The left-hand side is equal to

$$\frac{1}{k^\alpha}\int_0^1 \sum_{\substack{kx\leqq s_1<k(x+1)\\ 0\leqq s_2<k}} e^{2\pi i\lambda(s_1-s_2)}\,\varrho(\lambda)\,d\lambda =$$

$$= \frac{C}{k^\alpha}\int_0^1 e^{2\pi i\lambda kx} \left|\frac{e^{2\pi i\lambda k}-1}{e^{2\pi i\lambda}-1}\right|^2 \varrho(\lambda)\,d\lambda.$$

Let us change the variables in the last integral by putting $\lambda k=\mu$. We obtain

$$\frac{C}{k^\alpha}\int_0^1 e^{2\pi i\lambda kx}\left|\frac{e^{2\pi i\lambda k}-1}{e^{2\pi i\lambda}-1}\right|^2 \varrho(\lambda)\,d\lambda =$$

$$=\frac{C}{k^{\alpha+1}}\int_0^k e^{2\pi i\mu x}|e^{2\pi i\mu}-1|^2\sum_{m=-\infty}^{\infty}\frac{1}{|\mu/k+m|^{\alpha+1}}\,d\mu =$$

$$=C\int_0^k e^{2\pi i\mu x}|e^{2\pi i\mu}-1|^2\sum_{m=-\infty}^{\infty}\frac{1}{|\mu+mk|^{\alpha+1}}\,d\mu =$$

$$=C\sum_{l=0}^{k-1}\int_0^1 e^{2\pi i\mu x}|e^{2\pi i\mu}-1|^2\sum_{m=-\infty}^{\infty}\frac{1}{|\mu+l+km|^{\alpha+1}}\,d\mu =$$

$$=\int_0^1 e^{2\pi i\mu x}\varrho(\mu)\,d\mu.$$

Thus we get that a stationary Gaussian distribution with spectral density ϱ is scaling.

In order to prove that there are no other Gaussian scaling distributions it is sufficient to observe that the correlation $\langle\varphi(0),\varphi(0)\rangle$ and the scaling property together uniquely determine all $\langle\varphi(x),\varphi(0)\rangle$. Thus the scaling distributions form a one-parameter family. Since such a family has already been constructed the lemma is proved.

The following lemma is a natural generalization of Lemma 4.3.

Lemma 4.4. *Let $f(\lambda_1, \ldots, \lambda_d)$ be a positive function, homogeneous for $\lambda\neq 0$ of degree $d(\alpha+1)$. Then the spectral density*

(4.15)
$$\varrho(\lambda_1, \ldots, \lambda_d) = \prod_{s=1}^{d}|e^{2\pi i\lambda_s}-1|^2\sum_{m\in Z^d}\frac{1}{f(\lambda+m)}$$

corresponds to a Gaussian scaling distribution.

The proof of this lemma is a word-by-word repetition of the first part of the proof of the preceding lemma. But for $d>1$ we cannot state that formula (4.15) describes all possible Gaussian scaling distributions.

An important property of the expression for $\varrho(\lambda)$ is its integrable singularity at the point $\lambda=0$. The slow decrease of the correlation coefficients is provided by this singularity.

The Gaussian distributions, as seen in Chapter I, can be regarded as limit Gibbs distributions for quadratic Hamiltonians of the form $\mathscr{H}=\sum a(x-y)\varphi(x)\varphi(y)$, $a(x)=\int e^{2\pi i(\lambda,x)}\varrho^{-1}(\lambda)\,d\lambda$, where ϱ is the spectral density. Let us consider the function $f(\lambda_1, \ldots, \lambda_d)=\prod_{i=1}^{d}\lambda_i^2(\lambda_1^2+\ldots+\lambda_d^2)$ for $d>2$. Then $\varrho(\lambda)\sim\frac{\text{const}}{\lambda_1^2+\ldots+\lambda_d^2}$ for $\lambda\to 0$ and thus $\varrho^{-1}(\lambda)\sim\text{const}\cdot\sum_{i=1}^{d}\lambda_i^2$. It can be proved that $\varrho^{-1}(\lambda)$ is a real analytic function on the d-dimensional torus. This means that the interaction $a(x)$ corresponding to it decreases exponentially when $|x|\to\infty$, i.e. $a(x)$ is a short-range potential. The corresponding α is equal to $1+\frac{2}{d}$.

Now we can explain why the index η was written in the form given in Section 1. It is a widespread assumption that the scaling distributions appearing at β_{cr} as limit distributions for Gibbs distributions corresponding to Hamiltonians of finite interaction radius must be described by Hamiltonians with fast decreasing interaction. In the Gaussian case such Hamiltonians appear at $\alpha=1+\frac{2}{d}$ only. Therefore if $\eta\neq 0$, then the scaling distribution is non-Gaussian and the index η characterizes the degree of difference from the Gaussian scaling distribution.

7. The space of Hamiltonians and the definition of the linearized renormalization group

In this section we discuss an important aspect of the theory of scaling distributions and the renormalization group method: namely, the question of the stability of scaling probability distributions. The notion of stability has implicitly arisen in connection with Dyson's hierarchical models. First we discuss a more general problem, the problem of the space of Hamiltonians.

Let the variables $\varphi(x)$, $x\in Z^d$, take arbitrary real values. Let us consider the m-particle potential $\mathscr{U}(\varphi)=\sum c(x_1, \dots, x_m)\,\varphi(x_1)\dots\varphi(x_m)$ for every $m\geqq 1$. It is convenient to assume that all indices x_i run independently of each other through the lattice Z^d and $c(x_1, \dots, x_m)$ is a symmetric function of its arguments. For the time being we do not investigate the convergence of the series defining $\mathscr{U}$, but we assume that the decrease of the coefficients provides the convergence of the series $\sum_{x_1,\dots,x_m}|c(x_1, \dots, x_m)|$. Then to every potential $\mathscr{U}(\varphi)$ uniquely corresponds a Fourier series

$$u(\lambda)=\exp\{2\pi i\textstyle\sum_{r=1}^{m}(x_r, \lambda_r)\}\,c(x_1, \dots, x_m)$$

where $u(\lambda)$ is a continuous symmetric function defined on the md-dimensional torus; it has an absolutely convergent Fourier series. Let $\mathscr{A}^{(s)}(\mathrm{Tor}^{md})$ be the Banach space of these functions $u(\lambda)$, where the norm is defined by $\|u\|=\sum_{x_1,\dots,x_m}|c(x_1, \dots, x_m)|$. Thus we can consider the space of potentials $\mathscr{U}(\varphi)$ as a Banach space $\mathscr{A}^{(s)}(\mathrm{Tor}^{md})$. In addition, if $\mathscr{U}\in\mathscr{A}^{(s)}(\mathrm{Tor}^{m_1d})$, $\mathscr{V}\in\mathscr{A}^{(s)}(\mathrm{Tor}^{m_2d})$, then $\mathscr{U}\mathscr{V}\in\mathscr{A}^{(s)}(\mathrm{Tor}^{(m_1+m_2)d})$. Let $\mathscr{A}^{(s)}$ be the direct sum $\oplus\sum_{m\geqq 1}\mathscr{A}^{(s)}(\mathrm{Tor}^{md})$ of these spaces and let $\|\cdot\|_{\mathscr{A}^{(s)}}=\sum_{m\geqq 1}\|\cdot\|_{\mathscr{A}^{(s)}(\mathrm{Tor}^{md})}$ be the norm on it.

Every potential $\mathscr{U}\in\mathscr{A}^{(s)}(\mathrm{Tor}^{md})$ uniquely defines a translation-invariant Hamiltonian

$$\mathscr{H}(\varphi)=\sum_{t\in Z^d}\mathscr{U}(T^t\varphi)=\sum_{x_1,\dots,x_m\in Z^d}\sum_{x\in Z^d}c(x_1+x, \dots, x_m+x)\,\varphi(x_1)\dots\varphi(x_m).$$

The meaning of this series is only formal but the coefficient $d(x_1, ..., x_m) = \sum_{x \in Z^d} c(x_1 + x, ..., x_m + x)$ is uniquely determined if

$$\mathscr{U} \in \mathscr{A}^{(s)}(\mathrm{Tor}^{md}) \quad \text{and} \quad d(x_1+x, ..., x_m+x) = d(x_1, ..., x_m).$$

It is possible that two different potentials $\mathscr{U}_1, \mathscr{U}_2 \in \mathscr{A}^{(s)}(\mathrm{Tor}^{md})$ define the same Hamiltonian, i.e. the corresponding coefficients d coincide. Such potentials $\mathscr{U}_1, \mathscr{U}_2$ are said to be homologous.

Lemma 4.5. *Two potentials $\mathscr{U}_1$ and $\mathscr{U}_2$ are homologous if and only if the functions $u_1(\lambda)$, $u_2(\lambda)$ corresponding to them coincide on the $(m-1)d$-dimensional diagonals $\sum_{r=1}^m \lambda_r \equiv 0 \pmod 1$.*

Proof. The coefficients d can be written as

$$d(x_1, \ldots, x_m) = \int u(\lambda_1, \ldots, \lambda_m) e^{-2\pi i \sum_{p=1}^m (\lambda_p, x_p)} \delta\left(\sum_{p=1}^m \lambda_p\right) \prod_{p=1}^m d\lambda_p$$

where the symbol $\delta(\sum_{p=1}^m \lambda_p)$ means that the function $u(\lambda_1, \ldots, \lambda_m)$ is integrated over the $(m-1)d$-dimensional diagonals defined above. Hence it is concluded that the coefficients d determine uniquely the values of $u(\lambda)$ on the diagonals. The lemma is proved.

Now let us notice that the subset of the space $\mathfrak{A}^{(s)}(\mathrm{Tor}^{md})$, consisting of the functions $u(\lambda_1, \ldots, \lambda_m)$ which vanish on the $(m-1)d$-dimensional diagonals, forms a closed subspace $\mathscr{I}_0$. Therefore every m-particle Hamiltonian $\mathscr{H}$ can be regarded as a point of the factor space $\mathscr{A}^{(s)}(\mathrm{Tor}^{md})|\mathscr{I}_0 = \mathscr{B}^{(s)}(\mathrm{Tor}^{md})$. Thereby the space of m-particle Hamiltonians is a Banach space. Let us denote by $\mathscr{B}^{(s)}$ the direct sum $\oplus \sum_{m \geqq 1} \mathscr{B}^{(s)}(\mathrm{Tor}^{md})$ and let $\|\cdot\|_{\mathscr{B}^{(s)}} = \sum_{m \geqq 1} \|\cdot\|_{\mathscr{B}^{(s)}(\mathrm{Tor}^{md})}$ be the norm on $\mathscr{B}^{(s)}$.

Now we can define the linearized renormalization group. Let Q be a scaling distribution. For every function $f(\varphi) \in L^1(\mathfrak{M}, Q)$ and for every $k \geqq 1$ let us consider the conditional expectation

$$E(f(\varphi)|\mathfrak{A}_k \varphi) = (\partial \mathfrak{A}_k^*) f$$

which is a function of the sequence $\mathfrak{A}_k \varphi$. It is clear that $T^t \partial \mathfrak{A}_k^* = \partial \mathfrak{A}_k^* T^{tk}$ for every $t \in Z^d$. Let us assume that the scaling distribution Q fulfils the condition: $\partial \mathfrak{A}_k^*(U(\varphi)) \in \mathscr{A}^{(s)}$ for every potential $\mathscr{U}(\varphi) \in \mathscr{A}^{(s)}(\mathrm{Tor}^{md})$. Then we can write

$$(\partial \mathfrak{A}_k^*) \mathscr{H} = \sum_{t \in Z^d} (\partial \mathfrak{A}_k^*) \mathscr{U}(T^t \varphi)$$

for every Hamiltonian $\mathscr{H}(\varphi) = \sum_{t \in Z^d} u(T^t \varphi) \in \mathscr{B}^{(s)}$. It is clear that $\partial \mathfrak{A}_k^* \mathscr{H} \in \mathscr{B}^{(s)}$.

Definition 4.6. *The semigroup $\{\partial \mathfrak{A}_k^*\} = \partial \mathfrak{A}^*$ acting on the space $\mathscr{B}^{(s)}$ is called the linearized renormalization group.*

We explain the meaning of this name. Let us assume that a probability distribution Q_1 is absolutely continuous with respect to the scaling distribution Q. Let the derivative

$\frac{dQ_1}{dQ}$ be equal to $f(\varphi)$. Then it is easy to see that $\mathfrak{A}_k^* Q_1$ is absolutely continuous with respect to $\mathfrak{A}_k^* Q = Q$ and $\frac{d(\mathfrak{A}_k^* Q_1)}{dQ} = (\partial \mathfrak{A}_k^*) f$. From the point of view of the theory of limit Gibbs distributions, a translation-invariant distribution close to Q can be written as $Qe^{-\varepsilon\mathscr{H}}$, $\mathscr{H} \in \mathscr{B}^{(s)}$, and ε is a small parameter. Then, up to the higher order of ε, we formally have:

$$\mathfrak{A}_k^*(Qe^{-\varepsilon\mathscr{H}}) = \mathfrak{A}_k^*(Q(1-\varepsilon\mathscr{H}+\dots)) =$$
$$= Q(1-\varepsilon\partial\mathfrak{A}_k^*\mathscr{H}+\dots) = Qe^{-\varepsilon\partial\mathfrak{A}_k^*\mathscr{H}+\dots}$$

which explains the name "linearized renormalization group".

The Hamiltonian $\mathscr{H} \in \mathscr{B}^{(s)}$ is called an eigen-Hamiltonian (of the linearized renormalization group) iff $\partial\mathfrak{A}_k^*\mathscr{H} = k^\gamma \mathscr{H}$ for some γ. If $\gamma > 0$, $\gamma < 0$, $\gamma = 0$, then $\mathscr{H}$ is said to be unstable, stable, neutral, respectively. In the next section we investigate the action of the linearized renormalization group and the form of the eigen-Hamiltonians of the Gaussian scaling distributions.

8. The linearized renormalization group and its spectrum in the case of Gaussian scaling distributions

Let P be an arbitrary Gaussian distribution on the set $\mathfrak{M}$ with zero mean. Let us consider a product $\varphi(x_1)\dots\varphi(x_m)$, $x_p \in Z^d$, $1 \leq p \leq m$ (it belongs to the Hilbert space $\mathscr{L}^2(\mathfrak{M}, P)$).

Definition 4.7. *The Hermite–Ito polynomial of the random variable* $\varphi(x_1)\dots\varphi(x_m)$ *(further denoted by* $:\varphi(x_1)\dots\varphi(x_m):$*) is the perpendicular from the end of the vector* $\varphi(x_1)\dots\varphi(x_m)$ *to the closed subspace generated by all possible random variables* $\varphi(y_1)\dots\varphi(y_p)$

$$p < m, \quad y_i \in Z^d.$$

If $\mathscr{H} \in \mathscr{B}^{(s)}(\mathrm{Tor}^{md})$, $\mathscr{H} = \sum d(x_1, \dots, x_m)\varphi(x_1)\dots\varphi(x_m)$, *then the formal series*

$$:\mathscr{H}: = \sum d(x_1, \dots, x_m):\varphi(x_1)\dots\varphi(x_m):$$

is called the Hermite–Ito polynomial of the Hamiltonian $\mathscr{H}$. *Let us introduce the following notations:*

$$:\mathscr{A}^{(s)}(\mathrm{Tor}^{md}): = \{:\mathscr{U}(\varphi):,\ \mathscr{U}(\varphi) \in \mathscr{A}^{(s)}(\mathrm{Tor}^{md})\},$$

$$:\mathscr{B}^{(s)}(\mathrm{Tor}^{md}): = \{:\mathscr{H}(\varphi):,\ \mathscr{H} \in \mathscr{B}^{(s)}(\mathrm{Tor}^{md})\}.$$

The spaces $:\mathscr{A}^{(s)}(\mathrm{Tor}^{md}):$ and $:\mathscr{B}^{(s)}(\mathrm{Tor}^{md}):$ become Banach spaces provided that $\|:\mathscr{U}(\varphi):\|_{:\mathscr{A}^{(s)}(\mathrm{Tor}^{md}):}=\|\mathscr{U}_\varphi\|_{\mathscr{A}^{(s)}(\mathrm{Tor}^{md})}$ and $\|:\mathscr{H}(\varphi):\|_{:\mathscr{B}^{(s)}(\mathrm{Tor}^{md}):}=\|\mathscr{H}(\varphi)\|_{\mathscr{B}^{(s)}(\mathrm{Tor}^{md})}$. After this both Banach spaces $\oplus\sum_m :\mathscr{A}^{(s)}(\mathrm{Tor}^{md}):$ and $\oplus\sum_m :\mathscr{B}^{(s)}(\mathrm{Tor}^{md}):$ can be constructed.

The following lemma formulates an important property of the Hermite–Ito polynomials.

Lemma 4.6. *The Hermite–Ito polynomial* $:\varphi(x_1)\ldots\varphi(x_m):$ *is a polynomial of degree* m *of the variables* $\varphi(x_1), \ldots, \varphi(x_m)$ *with leading coefficient* 1 *and every one of its coefficients is uniformly bounded with respect to* $x_1, \ldots, x_m$.

Proof. We show that the Hermite–Ito polynomial $:\varphi(x_1)\ldots\varphi(x_m):$ is the perpendicular Θ dropped from the end of the vector $\varphi(x_1)\ldots\varphi(x_m)$ on the subspace generated by the random variables $\varphi(x_{j_1}), \ldots, \varphi(x_{j_p})$, $p<m$. We can write every random variable $\varphi(y)$, $y\neq x_1, \ldots, x_m$, as

$$\varphi(y) = \sum_{p=1}^{m} c(y, x_p)\varphi(x_p)+\psi(y) \tag{4.16}$$

where $\psi(y)\perp\varphi(x_p)$, $p=1, \ldots, m$. In the Gaussian case orthogonality is equivalent to independence. It is convenient to assume that if y coincides with one of the variables $x_1, \ldots, x_m$, then $c(y, x_p)=\delta(y, x_p)$ and $\psi(y)\equiv 0$. Every product $\varphi(y_1)\ldots\varphi(y_p)$ can be written as a linear combination of products $\psi(y_1)\ldots\psi(y_q)\varphi(y_{q+1})\ldots\varphi(y_p)$ where every y_r, $q+1\leqq r\leqq p$, coincides with one of the variables $x_1, \ldots, x_m$. If $q\neq 0$, then the orthogonality of Θ and the product $\psi(y_1)\ldots\psi(y_q)\cdot\varphi(y_{q+1})\ldots\varphi(y_q)$ follows from the independence of $\psi(y)$ and $\varphi(x_p)$, $1\leqq p\leqq m$. If $q=0$, then this statement follows from the definition of Θ. Thus

$$\Theta = :\varphi(x_1)\ldots\varphi(x_m):$$

The statement proved above can be interpreted as follows. Let us consider the density $\mathrm{const}\cdot\exp\left\{-\frac{1}{2}(Az, z)\right\}$ of the joint probability distribution of the random variables $\varphi(x_1), \ldots, \varphi(x_m)$ where $z=(z_1, \ldots, z_m)$ is an m-dimensional vector. Let $\mathscr{L}_A^2$ be the Hilbert space of all square-integrable (with respect to the above density) functions $f(z_1, \ldots, z_m)$.

Let $h(z_1, \ldots, z_m)$ be the orthogonal projection (in the sense of $\mathscr{L}_A^2$) of the product $z_1\ldots z_m$ on the subspace generated by all polynomials of degree less than m. The perpendicular $g(z_1, \ldots, z_m)=z_1\ldots z_m-h(z_1, \ldots, z_m)$ will be a polynomial of degree m of the variables $z_1, \ldots, z_m$ whose leading coefficient is equal to 1. On the other hand changing z_i to $\varphi(x_i)$, $1\leqq i\leqq m$ in $g(z_1, \ldots, z_m)$ we get $:\varphi(x_1)\ldots\varphi(x_m):$. The lemma is proved.

Lemma 4.7. *The following equalities hold*

1) $$\sum_{m'\leqq m} \mathscr{A}^{(s)}(\mathrm{Tor}^{m'd}) = \oplus \sum_{m'\leqq m} :\mathscr{A}^{(s)}(\mathrm{Tor}^{m'd}):;$$

2) $$\sum_{m'\leqq m} \mathscr{B}^{(s)}(\mathrm{Tor}^{m'd}) = \oplus \sum_{m'\leqq m} :\mathscr{B}^{(s)}(\mathrm{Tor}^{m'd}):.$$

Proof. Equalities 1) and 2) are obvious for $m=1$, since

$$\mathscr{A}^{(s)}(\mathrm{Tor}^{d}) =: \mathscr{A}^{(s)}(\mathrm{Tor}^{d}):; \quad \mathscr{B}^{(s)}(\mathrm{Tor}^{md}) =: \mathscr{B}^{(s)}(\mathrm{Tor}^{md}):.$$

For $m>1$ we prove them by induction. In virtue of Lemma 4.6 if $\mathscr{U}(\varphi)\in\mathscr{A}^{(s)}(\mathrm{Tor}^{md})$, then $:\mathscr{U}(\varphi):=c(x_1, \dots, x_m)\varphi(x_1)\dots\varphi(x_m)+\dots$ where dots refer to an element of the subspace $\oplus\sum_{m'<m}\mathscr{A}^{(s)}(\mathrm{Tor}^{m'd})$. By the induction hypothesis this element belongs to the subspace $\oplus\sum_{m'<m}:\mathscr{A}^{(s)}(\mathrm{Tor}^{m'd}):$. Thus $\mathscr{U}(\varphi)$ is an element of the space $\oplus\sum_{m'\leqq m}:\mathscr{A}^{(s)}(\mathrm{Tor}^{m'd}):$, hereby Equality 1) is proved. Equality 2) is a straightforward consequence of Equality 1). The lemma is proved.

The first information concerning the action of the linearized renormalization group is contained in the following lemma

Lemma 4.8. *For every k and m*

$$\partial\mathfrak{A}_k^*(:\mathscr{A}^{(s)}(\mathrm{Tor}^{md}):) = :\mathscr{A}^{(s)}(\mathrm{Tor}^{md}):$$
$$\partial\mathfrak{A}_k^*(:\mathscr{B}^{(s)}(\mathrm{Tor}^{md}):) = :\mathscr{B}^{(s)}(\mathrm{Tor}^{md}):.$$

Proof. The statement of the lemma can be obtained as a consequence of the general properties of Gaussian distributions, but we prefer to give an independent proof here. The inclusions

$$\partial\mathfrak{A}_k^*(:\mathscr{A}^{(s)}(\mathrm{Tor}^{md}):) \supset :\mathscr{A}^{(s)}(\mathrm{Tor}^{md}):,$$
$$\partial\mathfrak{A}_k^*(:\mathscr{B}^{(s)}(\mathrm{Tor}^{md}):) \supset :\mathscr{B}^{(s)}(\mathrm{Tor}^{md}):$$

are obvious; the differential $\partial\mathfrak{A}_k^*$ acts in the same way as the identical transformation does on the potentials $\mathscr{U}(\varphi)$ which in fact depend on the variables $\mathfrak{A}_k\varphi$. The proof of the reverse inclusion can be carried out by induction. The required equalities are obvious for $m=1$. Assuming that they are proved for every $m'<m$ we prove that they hold for m, too. Let us suppose that $:\mathscr{U}(\varphi):\in:\mathscr{A}^{(s)}(\mathrm{Tor}^{md}):$. Then we can write

$$\int \partial\mathfrak{A}_k^*(:\mathscr{U}(\varphi):)\cdot\mathscr{V}(\varphi)\,d(\mathfrak{A}_k^*P_0)(\varphi) =$$
$$= \int :\mathscr{U}(\varphi):\mathscr{V}(\varphi)\,dP_0(\varphi) = 0$$

for every $\mathscr{V}(\varphi)=\mathscr{V}(\mathfrak{A}_k(\varphi))\in:\mathscr{A}^{(s)}(\mathrm{Tor}^{m'd}):\ m'<m$ since $\mathscr{V}(\varphi)\in:\mathscr{A}^{(s)}(\mathrm{Tor}^{m'd}):$ and $:\mathscr{A}^{(s)}(\mathrm{Tor}^{m'd}):\perp:\mathscr{A}^{(s)}(\mathrm{Tor}^{md}):$ for $m'\neq m$. Thus $\partial\mathfrak{A}_k:\mathscr{U}(\varphi):\in:\mathscr{A}^{(s)}(\mathrm{Tor}^{md}):$. Thereby the first equality is proved. The second equality can be shown analogously. The lemma is proved.

Lemma 4.8 shows that the subspaces $:\mathscr{A}^{(s)}(\mathrm{Tor}^{md}):$, $:\mathscr{B}^{(s)}(\mathrm{Tor}^{md}):$ are invariant with respect to the linearized renormalization group.

Lemma 4.9. *Let G be a Gaussian stationary distribution with spectral density $\varrho_G(\lambda)$. Then the spectral density of the Gaussian stationary distribution $\mathfrak{A}_k^* G$ is of the form*

$$\varrho_{\mathfrak{A}_k^* G}(\lambda) = \frac{1}{k^{(\alpha+1)d}} \Pi_{r=1}^d |e^{2\pi i \lambda_r} - 1|^2 \cdot$$

$$\cdot \sum_{\substack{0 \leq p_s < k \\ 1 \leq s \leq d}} \varrho_G\left(\frac{\lambda_1}{k} + \frac{p_1}{k}, \ldots, \frac{\lambda_d}{k} + \frac{p_d}{k}\right) \Pi_{r=1}^d \left|e^{2\pi i \frac{\lambda_r + p_r}{k}} - 1\right|^{-2}.$$

Proof. Let us introduce the notation $\varphi_k(x) = \frac{1}{k^{d\alpha/2}} \sum_{y \in \Delta_k(x)} \varphi(y)$. Then

$$E(\varphi_k(x), \varphi_k(0)) =$$

$$= \frac{1}{k^{\alpha d}} \int e^{2\pi i k(x, \lambda)} \Pi_{r=1}^d \left| \frac{e^{2\pi i \lambda_r k} - 1}{e^{2\pi i \lambda_r} - 1} \right|^2 \varrho_G(\lambda) \, d\lambda.$$

Putting $k\lambda = \mu$ and splitting up the domain of integration into cubes we get the required formula. The lemma is proved.

Under the conditions of the preceding lemma we can obtain the formula

$$\varphi(x) = \sum c_k(x, y) \varphi_k(y) + \psi(x) \tag{4.17}$$

for every $x \in Z^d$ where the coefficients $c(x, y)$ are chosen so that the random variables $\psi(x)$ be independent of all random variables $\varphi_k(y)$. Another interpretation of formula (4.17) can be expressed in terms of Hamiltonians: if $\mathscr{H}(\varphi)$ is a Hamiltonian corresponding to the Gaussian distribution G, then $\mathscr{H}(\varphi) = \mathscr{H}_1(\mathfrak{A}_k \varphi) + \mathscr{H}(\psi)$. The random variables $\psi(x)$ possess the important property: $\sum_{y \in \Delta_k(x)} \psi(y) = 0$ for every x. Indeed let us take the values of the left-hand side and the right-hand side at all sites of an arbitrary cube $\Delta_k(x)$ and sum them up; then we get $k^{\alpha d/2} \varphi_k(x)$ on the left-hand side and a linear combination of φ_k plus $\sum_{y \in \Delta_k(x)} \psi(y)$ on the right-hand side. In other words $\sum_{y \in \Delta_k(x)} \psi(y)$ can be written as a linear combination of $\varphi_k(y)$. But from the independence of ψ and φ_k we can conclude that $\sum_{y \in \Delta_k(x)} \psi(y)$ is equal to 0. Now we can give an explicit formula for the coefficients $c_k(x, y)$. Put $d_k(x, y) = E\varphi(x) \varphi_k(y)$, $b(x) = b_{\mathfrak{A}_k^* G}(x) = E\varphi_k(x) \varphi_k(0)$. Multiplying by $\varphi_k(z)$ the left-hand side and the right-hand side of (4.17) and taking the expectation we get:

$$d_k(x, z) = \sum c_k(x, y) b(y - z).$$

Furthermore,

$$d_k(x, z) = \frac{1}{k^{d\alpha/2}} \int \exp\{2\pi i(\lambda, x)\} \sum_{\substack{kz_i \leq p_i < k(z_i+1) \\ i=1, \ldots, m}} \exp\{-2\pi i \textstyle\sum_{r=1}^d \lambda_r p_r\} \varrho_G(\lambda) \, d\lambda =$$

$$= \frac{1}{k^{d\alpha/2}} \int \exp\{2\pi i(\lambda, x - kz)\} \Pi_{r=1}^d \frac{e^{-2\pi i k \lambda_r} - 1}{e^{-2\pi i \lambda_r} - 1} \varrho_G(\lambda) \, d\lambda.$$

Therefore if $\|a(p'-p)\| = \|b(p'-p)\|^{-1}$, then

$$a(p) = \int \exp\{2\pi i(\mu, p)\}\, \varrho^{-1}_{\mathfrak{A}_k^* G}(\mu) =$$
$$= \int \exp\{2\pi i(k\lambda, p)\}\, \varrho^{-1}_{\mathfrak{A}_k^* G}(k\lambda),$$

where the domain of integration is the d-dimensional torus in both integrals. From the Parseval relation we get

$$c_k(x, y) = \sum d_k(x, z)\, a(z-y) =$$
$$= \frac{1}{k^{d\alpha/2}} \cdot \sum_{p\in Z^d} \int \exp\{2\pi i(\lambda, x-kp)\} \prod_{r=1}^d \frac{\exp\{-2\pi i k\lambda_r\}-1}{\exp\{-2\pi i\lambda_r\}-1}\, \varrho_G(\lambda)\, d\lambda \cdot$$
$$\cdot \int \exp\{2\pi i(\mu, p-y)\}\, \varrho^{-1}(\mu)\, d\mu =$$
$$= \frac{1}{k^{d\alpha/2}} \exp\{2\pi i(\lambda, x-ky)\} \cdot \prod_{r=1}^d \frac{(\exp(-2\pi i k\lambda_r)-1)}{\exp(-2\pi i\lambda_r)-1}\, \frac{\varrho_G(\lambda)\, d\lambda}{\varrho_{\mathfrak{A}_k^* G}(k\lambda)}.$$

Let $\varrho_G^{(1)}(\lambda)$ be equal to $\prod_{r=1}^d (\exp(-2\pi i\lambda_r)-1)^{-1} \varrho_G(\lambda)$ for an arbitrary Gaussian distribution G. Finally we can write

$$c_k(x, y) = \frac{1}{k^{d\alpha/2}} \int \exp\{2\pi i(\lambda, x-ky)\}\, \varrho_G^{(1)}(\lambda) \left(\varrho^{(1)}_{\mathfrak{A}_k^* G}(k\lambda)\right)^{-1} d\lambda. \tag{4.18}$$

Now we can directly investigate the spectrum of the linearized renormalization group for the Gaussian scaling distribution G.

Lemma 4.10. *Let* $:U:$ *be an element of* $:\mathscr{A}^{(s)}(\mathrm{Tor}^{md}):$ *and let* $u(\lambda_1, \ldots, \lambda_m)$ *be the Fourier transform of* U, *i.e. a symmetric (with respect to* $\lambda_1, \ldots, \lambda_m$*) function having an absolutely convergent Fourier series on the md-dimensional torus.*

Then $\partial\mathfrak{A}_k^*(:U:) = :V: \in :\mathscr{A}^{(s)}(\mathrm{Tor}^{md}):$ *and the Fourier transform* $v(\lambda_1, \ldots, \lambda_m)$ *of the potential* V *takes the form*

$$v(\lambda_1, \ldots, \lambda_m) = \frac{1}{k^{md(\alpha/2+1)}} \frac{1}{\prod_{r=1}^m \varrho_G^{(1)}(\lambda_r)} \sum_{\substack{0\le p_j<k \\ 1\le j\le m}} \cdot$$
$$\cdot \prod_{j=1}^m \varrho_G^{(1)}\left(\frac{\lambda_j+p_j}{k}\right) u\left(\frac{\lambda_1+p_1}{k} + \ldots + \frac{\lambda_m+p_m}{k}\right).$$

Proof. Let the equalities $U(\varphi) = \sum c(x_1, \ldots, x_m)\varphi(x_1) \ldots \varphi(x_m)$, $:U(\varphi): = \sum c(x_1, \ldots, x_m)\varphi(x_1) \ldots \varphi(x_m) + \ldots$, hold where dots refer to terms of degree less than m. To find $\partial\mathfrak{A}_k^* U(\partial\mathfrak{A}_k^*(:U:))$ it is sufficient to consider (4.17) applied for $U(\varphi)$ rather than φ and integrate the result over all $\psi(y)$. It is clear that in this way we get

$$:V: = \sum c'(x_1, \ldots, x_m)\, \varphi_k(x_1) \ldots \varphi_k(x_m) + \ldots,$$
$$c'(x_1, \ldots, x_m) = \sum_{y_1, \ldots, y_m \in Z^d} c_k(y_1, x_1) \ldots c_k(y_m, x_m)\, c(y_1, \ldots, y_m).$$

Since $c(y_1, \ldots, y_m)=\int \exp\{-2\pi i \sum_{p=1}^{m}(\lambda_p, y_p)\}u(\lambda_1, \ldots, \lambda_m)d\lambda$ we get, using the expression (4.18) for the coefficients $c_k(y, x)$,

$$\sum_{y_1,\ldots,y_m\in Z^d} c_k(y_1, x_1)\ldots c_k(y_m, x_m)c(y_1, \ldots, y_m) =$$

$$= \frac{1}{k^{dm\alpha/2}}\sum_{y_1,\ldots,y_m\in Z^d}\int u(\lambda_1, \ldots, \lambda_m)\,d\lambda\,\exp\{-2\pi i\sum_{p=1}^{m}(\lambda_p, y_p)\}\cdot$$

$$\cdot\prod_{p=1}^{m}\int \exp\{2\pi i(\mu_p, y_p-kx_p)\}\varrho_G^{(1)}(\mu_p)(\varrho_G^{(1)}(k\mu_p))^{-1}d\mu_p = \frac{1}{k^{dm\alpha/2}}\cdot$$

$$\cdot\int u(\lambda_1, \ldots, \lambda_m)\,d\lambda\int\sum_{y_1,\ldots,y_m\in Z^d}\exp\{-2\pi i\sum_{p=1}^{m}((\lambda_p-\mu_p), y_p)\}\cdot$$

$$\cdot\exp\{-2\pi ik\sum_{p=1}^{m}(x_p, \mu_p)\}\prod_{p=1}^{m}\varrho_G^{(1)}(\mu_p)(\varrho_G^{(1)}(k\mu_p))^{-1}\,d\mu =$$

$$= \frac{1}{k^{dm\alpha/2}}\int\exp\{-2\pi ik\sum_{p=1}^{m}(x_p, \lambda_p)\}u(\lambda_1, \ldots, \lambda_m)\cdot\prod_{p=1}^{m}\varrho_G^{(1)}(\lambda_p)(\varrho_G^{(1)}(k\lambda_p))^{-1}\,d\lambda.$$

Then putting $k\lambda_p=\mu_p$ we get the required formula. The lemma is proved.

It follows from Lemma 4.9 that if, for every y, $\sum_x |c_k(y, x)|<\text{const}<\infty$, then $V\in\mathscr{A}^{(s)}(\text{Tor}^{md})$. We prove this statement in the special case when the function $f(\lambda)$ in (4.15) is of the form $f(\lambda)=\text{const}\cdot\prod_{p=1}^{d}\lambda_p^2|\lambda|^{(\alpha-1)d}$.

This assumption means that the correlation function is asymptotically isotropic.

Since $c_k(y, x)=c_k(y-kx)$ are the Fourier coefficients of the function $\chi(\lambda)=\varrho_G^{(1)}(\lambda)(\varrho_G^{(1)}(k\lambda))^{-1}$ it is sufficient to analyse how the Fourier coefficients of the function $\chi(\lambda)$ decrease. It is clear that $\chi(\lambda)$ is smooth except in the points $\lambda=\left\{\frac{p_j}{k}, 1\leq j\leq d\right\}$. Let us first investigate the neighbourhood of the point 0. It is easy to see that in the neighbourhood of 0

$$\varrho_G^{(1)}(\lambda) = g_1(\lambda)\left(\frac{1}{\prod_{j=1}^{d}\lambda_j|\lambda|^{(\alpha-1)d}}+g_2(\lambda)\prod_{j=1}^{d}\lambda_j\right),$$

where g_1, g_2 are infinitely differentiable in the neighbourhood of 0 and $g_1(0)\neq 0$, $g_2(0)\neq 0$. Therefore in the neighbourhood of 0

$$\varrho_G^{(1)}(\lambda)(\varrho_G^{(1)}(k\lambda))^{-1} = \text{const}\cdot g_1(\lambda)(g_1(k\lambda))^{-1}\frac{1+\prod_{j=1}^{d}\lambda_j^2|\lambda|^{(\alpha-1)d}g_2(\lambda)}{1+A\prod_{j=1}^{d}\lambda_j^2|\lambda|^{(\alpha-1)d}g_2(k\lambda)}$$

for a constant A, $A\neq 0, 1$. In other words, in the neighbourhood of 0

$$\chi(\lambda) = h_1(\lambda)+\prod_{j=1}^{d}\lambda_j^2|\lambda|^{(\alpha-1)d}h_2(\lambda)$$

where h_1, h_2 are infinitely differentiable functions and different from 0. Analogously we get $\chi(\lambda)-\chi(\lambda_p)\sim\text{const}\cdot\prod_{j=1}^{k}\left(\lambda_j-\frac{p_j}{k}\right)|\lambda-\lambda_p|^{(\alpha-1)d}$ in the neighbourhoods of the points $\lambda_p=\left\{\frac{p_j}{k}\right\}_{j=1}^{d}$.

Such singularities involve, as well known, that the Fourier coefficients decrease as const $\cdot |x|^{-\alpha d}$ (notice that $1<\alpha<2$). The required result follows from this last observation.

In what follows we consider the only case when $f(\lambda)=c\prod_{j=1}^{d}\lambda_j^2|\lambda|^{(\alpha-1)d}$ in (4.15). It is implied from the statement proved above that $\partial\mathfrak{A}_k^*(:\mathscr{A}^{(s)}(\mathrm{Tor}^{md}):)=:\mathscr{A}^{(s)}(\mathrm{Tor}^{md}):$ and therefore $\partial\mathfrak{A}_k^*(:\mathscr{B}^{(s)}(\mathrm{Tor}^{md}):)=:\mathscr{B}^{(s)}(\mathrm{Tor}^{md}):$.

Let $h(\lambda)$ be an arbitrary homogeneous function of degree γ. Let us define the functions $\varrho_h^{(2)}(\lambda;\tau)$ and $\varrho_h^{(3)}(\lambda;\tau)$ as follows:

$$\varrho_h^{(2)}(\lambda;\tau)=\sum_{m\in Z^d}\frac{1}{h(\lambda+m)}\exp\{2\pi i(\tau,\lambda+m)\},\quad \tau\in R^d$$

$$\varrho_h^{(3)}(\lambda;\tau)=\varrho_h^{(2)}(\lambda;\tau)(\varrho_G^{(1)}(\lambda))^{-1}.$$

Let us suppose that $\varrho_h^{(3)}(\lambda;\tau)\in\mathscr{A}^{(s)}(\mathrm{Tor}^d)=\mathscr{A}(\mathrm{Tor}^d)$ for every $\tau\in R^d$. Let us consider the Gaussian random process $\zeta_h(\varphi;\tau)=\sum\Gamma(x;\tau)\varphi(x)$ with continuous time parameter where $\Gamma(x;\tau)=\Gamma(x-\tau)=\int\exp\{-2\pi i(x,\lambda)\}\varrho_h^{(3)}(\lambda;\tau)d\lambda$. Then $\zeta_h(\varphi;\tau)\in \in\mathscr{A}^{(s)}(\mathrm{Tor}^d)=:\mathscr{A}^{(s)}(\mathrm{Tor}^d):$ and, by Lemma 4.10, $\partial\mathfrak{A}_k^*(\zeta_h(\varphi;\tau))$ has the Fourier transform

$$\frac{1}{k^{d(\alpha/2+1)}}(\varrho_G^{(1)}(\lambda))^{-1}\sum_{\substack{0\le p_j<k\\1\le j\le d}}\varrho_G^{(1)}\left(\frac{\lambda_j+p_j}{k}\right)\varrho_h^{(3)}\left(\frac{\lambda_j+p_j}{k}\right)=$$

$$=\frac{1}{k^{d(\alpha/2+1)-\gamma}}\varrho_h^{(3)}(\lambda).$$

This last equality can be written as $\partial\mathfrak{A}_*^k(\zeta_h(\varphi;\tau))=k^{\gamma-d(\alpha/2+1)}\zeta_h\left(\mathfrak{A}_k\varphi;\frac{\tau}{k}\right)$. Therefore

$\partial\mathfrak{A}_k^*[:\zeta_h^m(\varphi;\tau):]=k^{m(\gamma-d(\alpha/2+1))}:\zeta_h^m\left(\mathfrak{A}_k\varphi;\frac{\tau}{k}\right):$ for every $m>0$. Now we can construct the eigen-Hamiltonians of the linearized renormalization group.

Lemma 4.11. *The Hamiltonian* $\int_{R^d}:[\zeta_h(\varphi;\tau)]^m:d\tau$ *is an eigen-Hamiltonian of the linearized renormalization group.*

Proof. We have

$$\partial\mathfrak{A}_k^*\int_{R^d}:[\zeta_h(\varphi;\tau)]^m:d\tau=\int_{R^d}\partial\mathfrak{A}_k^*(:[\zeta_h(\varphi;\tau)]^m:)\,d\tau=$$

$$=k^{m(\gamma-d(\alpha/2+1))}\int_{R^d}:\left[\zeta_h\left(\mathfrak{A}_k\varphi;\frac{\tau}{k}\right)\right]^m:d\tau.$$

Let us change the variable in the last integral putting $\tau'=\frac{\tau}{k}$. Then we get

$$\partial\mathfrak{A}_k^*\int_{R^d}:[\zeta_h(\varphi;\tau)]^m:d\tau=k^{m(\gamma-d(\alpha/2+1))+d}\int_{R^d}:[\zeta_h(\mathfrak{A}_k\varphi;\tau')]^m:d\tau'.$$

The lemma is proved.

Let us look for functions h for which the function $\varrho_h^{(3)}(\lambda, \tau)$ has an absolutely convergent Fourier series. It is clear that the inequality $\gamma \leqq \alpha d$ is a necessary condition for this. It can be shown that it is sufficient, too. As for $\gamma = \alpha d$ and $h = f(\lambda)(\prod_{r=1}^{m} \lambda_r)^{-1}$ this statement can be easily checked. The corresponding eigenvalue is $\gamma_m = \left[m\left(\frac{\alpha}{2}-1\right)+1\right] d$.

The last consideration can be slightly generalized.

Let λ_j be a d-dimensional vector, and let $h(\lambda_1, ..., \lambda_m)$ be a homogeneous function (depending on dm variables) of degree γ. Set

$$\varrho^{(2)}(\lambda_1 \dots \lambda_m; \tau) = \sum_{p_1, \dots, p_m \in Z^d} \frac{1}{h(\lambda_1 + p_1, \dots, \lambda_m + p_m)} \cdot$$

$$\cdot \exp\{2\pi i \sum_{r=1}^{m} (\tau_r, \lambda_r + p_r)\}, \quad \tau = (\tau_1, \dots, \tau_m) \in R^{md}.$$

Further set

$$\varrho_h^{(3)}(\lambda_1, \dots, \lambda_m; \tau) = \varrho^{(2)}(\lambda_1, \dots, \lambda_m; \tau) \prod_{j=1}^{m} (\varrho_G^{(1)}(\lambda_j))^{-1}$$

and

$$\Gamma(x; \tau) = \Gamma(x - \tau) = \int \exp\{-2\pi i \sum_{r=1}^{m} (x_r, \lambda_r)\} \varrho_h^{(3)}(\lambda_1, \dots, \lambda_m; \tau)\, d\lambda_1 \dots d\lambda_m,$$

$$x = (x_1 \dots x_m), \quad x_l \in Z^d, \quad 1 \leqq l \leqq m,$$

and construct a dm-dimensional field $\zeta_h(\varphi; \tau) = \sum \Gamma(x-\tau) :\varphi(x_1) \dots \varphi(x_m):$. Then the same consideration as above shows that $\partial \mathfrak{A}_k^*(\zeta_h(\varphi; \tau)) = k^{\gamma - md(\alpha/2+1)} \zeta_h(\mathfrak{A}_k \varphi; \tau)$ and $\int_{R^{md}} \zeta_h(\varphi; \tau)\, d\tau$ is an eigen-Hamiltonian with the eigenvalue $\gamma_h = \gamma - md\left(\frac{\alpha}{2}+1\right) + d$.

Discussion of the results. It seems that not all eigen-Hamiltonians constructed above are essential. To discuss this question let us first consider the case when $(\alpha - 1)d$ is not an even integer. The Hamiltonian corresponding to the Gaussian scaling distribution is of the form

$$\mathscr{H} = \sum a(x-y) \varphi(x) \varphi(y), \quad a(x) \sim \frac{\text{const}}{|x|^{\alpha d}},$$

i.e. it is generated by a long-range potential. The linearized renormalization group is a tool for analysing the corrections to a Hamiltonian. Therefore the assumption that it suffices to consider eigen-Hamiltonians with interaction decreasing at least as fast as $\frac{1}{|x|^{\alpha d+1}}$ is reasonable.

If one assumes that the interaction can be expanded in powers of $\frac{1}{|x|^{\alpha d+p}}$ (p is an integer), then the order of homogeneity of the function $h(\lambda_1, \dots, \lambda_m)$ appearing in the construction of eigen-Hamiltonians takes the values $\gamma = m\alpha, m\alpha - 2, \dots$. Then we get the eigenvalues $\gamma_m = \left[m\left(\frac{\alpha}{2}-1\right)+1\right]d$, $\gamma_m = \left[m\left(\frac{\alpha}{2}-1\right)+1\right]d-2$ etc.

Now let us note that we are dealing with the analysis of a ferromagnetic cricital point, so it suffices to restrict ourselves to even Hamiltonians; thus $m = 2m_1$. Now we have $\gamma_2 = (\alpha-1)\,d > 0$, i.e. there always is an unstable eigen-Hamiltonian in the space of quadratic forms. Since $\gamma_4 = 2\alpha - 3$, it is obvious that $\gamma_4 \leqq 0$ for $\alpha \leqq \frac{3}{2}$. Consequently if $\alpha \leqq \frac{3}{2}$, then γ_2 is the only non-negative eigenvalue and so we get a necessary condition $\alpha \leqq \frac{3}{2}$ for the thermodynamic stability of the Gaussian distributions. Let us notice that the values $\frac{\gamma_m}{d} = m\left(\frac{\alpha}{2}-1\right)+1$ become the logarithms of eigenvalues for the Gaussian solution of the hierarchical model by putting $\alpha = 2 - \log_2 c$, $k = 2$.

Another necessary condition can be obtained as follows. Let us consider a homogeneous function with degree of homogeneity $\gamma = d\alpha - 2$ for $m = 2$, i.e. in the space of quadratic forms. Then the eigenvalue is equal to $(\alpha-1)d-2$ and we get the abovementioned necessary condition for the stability of the Gaussian distribution: $(\alpha-1)d-2 \leqq 0$, i.e. $\alpha \leqq 1 + \frac{2}{d}$.

It is convenient to represent both conditions on a diagram (see Figure 4.3).

If the number $(\alpha-1)d$ is integer, then the Hamiltonian of the corresponding Gaussian scaling distribution is generated by an exponentially decreasing (i.e. shortrange) potential. Hence a real analytic function defined on the md-dimensional

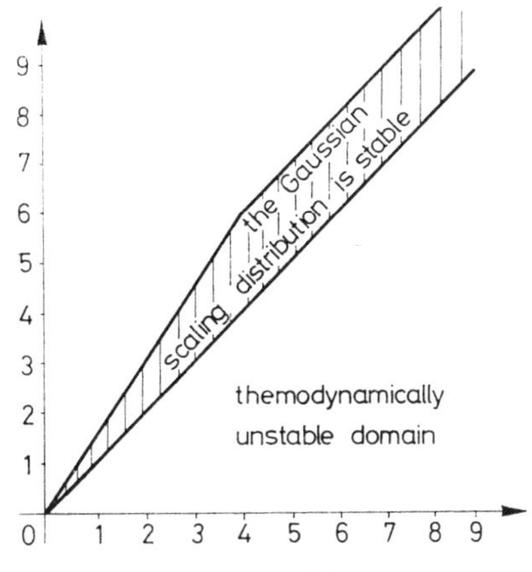

FIGURE 4.3

torus corresponds to this Gaussian distribution. Thus it is reasonable to consider only such eigen-Hamiltonians which correspond to some real analytic functions. It is easy to check that in this case there are no new conditions on the stability of the Gaussian distributions.

9. Bifurcation points, non-Gaussian scaling distributions, ε-expansions

The previous analysis shows that if $\alpha_m=2-\frac{1}{m}$, $m=1, 2, 3 \ldots$, then there are neutral eigen-Hamiltonians. It is natural to expect that similarly to the case of the hierarchical model, a new, non-Gaussian branch of scaling distribution appears in the neighbourhood of every point α_m. In this section the formal ε-expansion of the Hamiltonians of these distributions is constructed. For the sake of simplicity we consider the case $m=1$, $\varepsilon=\alpha-\frac{3}{2}$.

Let $\mathscr{H}_0(\varphi)$ be the Hamiltonian of the Gaussian scaling distribution for a given α. We look for the Hamiltonian $\mathscr{H}(\varphi)$ of the unknown non-Gaussian scaling distribution by writing

$$\mathscr{H}(\varphi) = \mathscr{H}_0(\varphi)+\varepsilon\mathscr{H}_1(\varphi)+\varepsilon^2\mathscr{H}_2(\varphi)+\ldots,$$

where $\mathscr{H}_i\in\mathscr{B}^{(s)}$, $i=1, 2, \ldots$. Let us simplify our considerations by assuming that $\mathscr{H}(\varphi)$ is invariant under the action of the semigroup $\mathfrak{A}^{(0)}$ generated by the group of space translations together with the cyclic subsemigroup $\{\mathfrak{A}_{2^k}\}$, whereas invariance with respect to the whole semigroup $\mathfrak{A}_0$ is not supposed. The reason for this simplification is the following: we can look for the translation-invariant Hamiltonian $\mathscr{H}(\varphi)$ by starting from the condition of its invariance with respect to the unique transformation $\mathfrak{A}_2^*$ only. Let us write (4.17) in the abbreviated form $\varphi=C_2\varphi_2+\psi$. The operator $\mathfrak{A}_2^*$ acts on $\mathscr{H}(\varphi)$ by changing φ to $C_2\varphi_2+\psi$ and integrating with respect to ψ. The condition of the scaling property can formally be written as

$$e^{-\mathscr{H}(\varphi_2)} = \int e^{-\mathscr{H}(c_2\varphi_2+\psi)}\,d\psi.$$

Further on, in virtue of the relation $\mathscr{H}_0(\varphi_2)+\mathscr{H}_0(\psi)=\mathscr{H}_0(c_2\varphi_2+\psi)$, we have the equality

$$e^{-\varepsilon\mathscr{H}_1-\varepsilon^2\mathscr{H}_2-\ldots} = \int e^{-\mathscr{H}_0(\psi)-\varepsilon\mathscr{H}_1(c_2\varphi_2+\psi)-\varepsilon^2\mathscr{H}_2(c_2\varphi_2+\psi)-\ldots}\,d\psi \tag{4.19}$$

for the correction $\varepsilon\mathscr{H}_1+\varepsilon^2\mathscr{H}_2+\ldots$.

This relation is analogous to the basic integral equation of the theory of hierarchical models.

The functional integral on the right-hand side will be treated by means of perturbation theory.

We assume that the correction $\varepsilon\mathscr{H}_1+\varepsilon^2\mathscr{H}_2$ has the form

$$(\varepsilon_1 a_1+\varepsilon^2 a_2+\varepsilon^3 a_3+\dots)\, G_4+\varepsilon^2 I_2+\varepsilon^3 I_3+\varepsilon^4 I_4+\dots$$

where $a_1, a_2, a_3 \dots$ are numbers, independent of ε, $2\alpha-3=2\varepsilon$ is an eigenvalue of the linearized renormalization group, the corresponding eigen-Hamiltonian is $G_4\in :\mathscr{B}^{(s)}(\mathrm{Tor}^{md}):$, and $I_j\in\mathscr{B}^{(s)}$, $j=2, 3, \dots$, are some correction Hamiltonians also independent of ε. We show how to find a_1 and I_2. Let us formally expand the right-hand side of (4.19) in terms of ε up to the term of second order, and then integrate it:

$$\int e^{-\mathscr{H}_0(\psi)-(\varepsilon a_1+\varepsilon^2 a_2+\dots)G_4-\varepsilon^2 I_2-\varepsilon^3 I_3-\varepsilon^4 I_4-\dots}\, d\psi =$$

$$=\int e^{-\mathscr{H}_0(\psi)}\, d\psi \left(1-(\varepsilon a_1+\varepsilon^2 a_2+\dots)G_4+\frac{\varepsilon^2}{2} a_1^2 G_4^2-\varepsilon^2 I_2+\dots\right) d\psi =$$

$$=1-2^{2\varepsilon}(\varepsilon a_1+\varepsilon^2 a_2+\dots)\, G_4+\frac{\varepsilon^2}{2} a_1^2\partial\mathfrak{A}_2^*(G_4^2)-\varepsilon^2\partial\mathfrak{A}_2^*(I_2)+\dots.$$

The function G_4^2 does not belong to the space $\mathscr{B}^{(s)}$ but the action of $\partial\mathfrak{A}_2^*$ on G_4^2 can be defined term by term in a reasonable way. The result can be written as

$$\partial\mathfrak{A}_2^*\, G_4^2 = G_4^2+J_2+J_4+J_6$$

where $J_{2j}\in :\mathscr{B}^{(s)}(\mathrm{Tor}^{2jd}):$, $j=1, 2, 3$. Expanding the left-hand side of (4.19) into a series with respect to ε up to the term of second order we obtain

$$e^{-\varepsilon\mathscr{H}_1-\varepsilon^2\mathscr{H}_2-\dots} = e^{-(\varepsilon a_1+\varepsilon^2 a_2+\dots)G_4-\varepsilon^2 I_2-\dots} =$$

$$=1-\varepsilon a_1 G_4+\frac{\varepsilon^2}{2} a_1^2 G_4^2-\varepsilon^2 a_2 G_4-\varepsilon^2 I_2+\dots$$

Comparison of the first-order terms leads to an identity, but a non-trivial equation can be deduced by comparing the terms of second order:

$$-a_2 G_4-I_2 = -2(\ln 2)\, a_1 G_4-a_2 G_4+J_2+J_4+J_6-\partial\mathfrak{A}_2^*\, I_2$$

or

$$\partial\mathfrak{A}_2^*\, I_2-I_2 = -2\ln 2a_1+J_2+J_4+J_6. \tag{4.20}$$

Since every subspace $:\mathscr{B}^{(s)}(\mathrm{Tor}^{md}):$ is invariant under $\partial\mathfrak{A}_2^*$, the last equation can separately be solved in each of them. Let $I_{2,j}$ be the projection of I_2 on $:\mathscr{B}^{(s)}(\mathrm{Tor}^{2jd}):$. Then (4.20) is equivalent to the following three equations:

$$\partial\mathfrak{A}_2^*\, I_{2,2}-I_{2,2} = J_2$$

$$\partial\mathfrak{A}_2^*\, I_{2,4}-I_{2,4} = -2\ln 2a_1+J_4$$

$$\partial\mathfrak{A}_2^*\, I_{2,6}-I_{2,6} = I_6.$$

In the first and third spaces the spectrum is uniformly separated from 1, so we can write

$$I_{2,2} = (\partial\mathfrak{A}_2^* - Id)^{-1}J_2, \quad I_{2,6} = (\partial\mathfrak{A}_2^* - Id)^{-1}J_6$$

where Id is the identity operator. In the subspace $:\mathscr{B}^{(s)}(\mathrm{Tor}^{4d})$: the operator $(\partial\mathfrak{A}_2^* - Id)^{-1}$ has an eigenvalue $(2^{2\varepsilon}-1)^{-1} \sim \frac{1}{(2\ln 2)\varepsilon}$. In order to obtain a solution independent of ε it is necessary that the right-hand side have a projection equal to zero on the corresponding eigenvector. The coefficient a_1 can be determined by this condition. Namely, if the function $\beta(\lambda_1, \lambda_2, \lambda_3, \lambda_4)$ (possessing absolutely convergent Fourier series) corresponds to J_4, then for $a_1 = \frac{\beta(0, 0, 0, 0)}{2\ln 2}$ this projection is indeed equal to 0. After this we write $I_{2,4}$ as

$$I_{2,4} = (\partial\mathfrak{A}_2^* - Id)^{-1}(-(2\ln 2)a_1 + J_4)$$

and the solution is bounded for small ε'. P. M. Bleher has proved that $a_1 \neq 0$.

The method for determining further coefficients can be described by means of a diagram. We briefly present the corresponding considerations. Let us assume that the coefficients $a_1, a_2, \ldots, a_{m-2}$ and the Hamiltonians $I_2, \ldots, I_{m-1}$ have already been found. Let us expand the exponential function in (4.19) into a series with respect to ε up to the terms of the mth order. Let us write the terms of the mth order, and the term of $(m-1)$st order containing a_{m-1}:

$$\varepsilon^{m-1} a_{m-1} G_4 + \varepsilon^m \sum \frac{(-1)^{l_1+\ldots+l_{m-1}}}{l_1! \, l_2! \ldots l_{m-1}!} a_1^{l_1} (G_4)^{l_1} \prod_{j=2}^{m-1} (I_j + a_j G_4)^{l_j} - \varepsilon^m q_m.$$

To apply the transformation $\partial\mathfrak{A}_2^*$ to every term it is necessary to integrate with respect to ψ. We need to integrate two types of products of the variables ψ. The decomposable products (there is a factor which does not contain $\psi(x)$) and the non-decomposable products (every factor contains at least one variable $\psi(y)$). Let p_m be the result of the integration of non-decomposable products. The equation for q_m is of the form

$$a_{m-1}\varepsilon^{m-1}(2^{2\varepsilon}-1)G_4 + \varepsilon^m J_m = \varepsilon^m(\partial\mathfrak{A}_2^* I_m - I_m).$$

Let us write J_m as

$$J_m = \sum c_l^{(m)} G_{2l} + \tilde{J}_m$$

where all functions in the Fourier representation of $\tilde{J}_m$ are equal to 0 for $\lambda = 0$, and G_{2l} are the eigen-Hamiltonians of the Gaussian scaling distribution.

Set $a_{m-1} = -c_2^{(m)}(2\ln 2)^{-1}$. We look for I_m assuming that $I_m = \sum d_l^{(m)} G_{2l} \tilde{J} +_m$, $d_l^{(m)} = c_l^{(m)}(2^{\gamma_{2l}}-1)^{-1}$. We get

$$\tilde{I}_m = -\sum_{l=0}^{\infty} (\partial\mathfrak{A}_2^*)^l \tilde{J}_m$$

which is the required expression.

A number of problems arise in connection with the ε-expansion constructed above. Next we enumerate them.

1. The analytic nature of the ε-expansion. The process of finding the coefficients of $\tilde{J}_m$ described above has a formal character. It is not clear whether the coefficients of $\tilde{J}_m$ and I_m belong to the original space B. Besides, the coefficients of I_m are only determined for the cyclic subsemigroup $\{\mathfrak{A}^*_{2^k}\}_{k=0}^{\infty}$ but they should be determined for the whole semigroup $\{\mathfrak{A}_n\}$. These problems can be explained by developing the diagram technique used for determining the coefficients of I_m and by the analysis of the structure of these diagrams.

Another question is the behaviour of these series in the complex domain of the variable ε; in particular, it would be necessary to clarify the analytic nature of the asymptotic series with respect to ε for the critical indices corresponding to the non-Gaussian scaling distributions constructed in this way.

2. The domain of attraction of scaling distributions. As was mentioned, one of the basic problems of the theory of scaling distributions is to find the form of the Hamiltonians $\mathscr{H}$ for which a given scaling distribution appears as a limit distribution of the sums $\frac{1}{k^{d\alpha/2}}\sum_{y\in\Delta_k(x)}\varphi(y)$ for $k\to\infty$ (see Section 5) which is determined by starting from the limit Gibbs distribution for the Hamiltonian $\beta_{cr}\mathscr{H}$. In the domain of the parameters α, d where the Gaussian distribution is stable (see Figure 4.3), it is natural to expect that the derived Hamiltonians are local perturbations of quadratic Hamiltonians $\sum U(x_1-x_2)\varphi(x_1)\varphi(x_2)$, where $U(x)\sim\frac{f\left(\frac{x}{\|x\|}\right)}{\|x\|^{\alpha d}}$ and f is a continuous positive function on the unit sphere which takes into account the anisotropy of the interaction. Such Hamiltonians will be non-Gaussian since $\varphi(x)$ takes the values ± 1 or, in a slightly more general case, a finite number of values.

In the case of non-Gaussian scaling distributions constructed by means of ε-expansion, the situation will not be as clear as it was for the Gaussian case. The formal perturbation theory with respect to the parameter ε shows that systems with Hamiltonians having binary interaction decreasing at infinity as $U(x)\sim\frac{f\left(\frac{x}{\|x\|}\right)}{\|x\|^{\alpha d}}$ belong to the domain of attraction of these non-Gaussian scaling distributions up to any order furnished by perturbation theory. However, because of the asymptotic character of the series of perturbation theory we cannot exclude the possibility that an exponentially small (with respect to ε) renormalization of the exponent of the potential takes place. In my opinion no renormalization of the exponent happens, but the situation as a whole is not clear. It can be hoped that these questions will be clarified by generalizing and developing the technique used for the construction of non-Gaussian solutions for hierarchical models.

3. Let us assume that there is no additional renormalization of the exponent of the potential. Then the branch of non-Gaussian scaling distributions constructed above describes the behaviour of those systems in the neighbourhood of the point β_{cr} whose interaction at long distances is binary and the potential of the binary interaction decreases as $\frac{1}{r^{d(3/2+\varepsilon)}}$ for $d=1, 2, 3$. In other words the critical indices depend on ε, i.e. on the asymptotic behaviour of the decrease of the interaction potential, if ε is small. It is reasonable to assume (for every dimension d) the existence of a number $\varepsilon^*(d)$ such that for $\varepsilon \geqq \varepsilon^*(d)$ the critical indices do not depend on the exponent of the potential, and that they are the same as those of the short-range potentials. In other words there exists for every dimension an exponent $\gamma^*(d)=$ $=d\left(\frac{3}{2}+\varepsilon^*\right)$ of the potential such that the systems with interaction potential of exponent $\gamma \geqq \gamma^*(d)$ behave as short-range systems from the point of view of the theory of phase transitions of the second kind. It is not excluded that $\varepsilon^*(d)=\frac{4-2d}{2d}$ since then the formal series of the non-Gaussian scaling distribution would contain Hamiltonians with exponentially decreasing interaction. If this series were convergent the Hamiltonian corresponding to it would be a Hamiltonian with rapidly decreasing interaction and the latter would generate a scaling distribution appearing at β_{cr} for the systems with short-range potential. However, for $d=2$ we get $\alpha^*=2$ for which the series is divergent according to the well-known result on the two-dimensional Ising model, which follows from Onsager's solution.

There are other logically possible cases, e.g. the point $\varepsilon^*(d)$ can be a point of strict loss of stability. This means that for $\varepsilon<\varepsilon^*(d)$ the scaling distributions belonging to the branch constructed above appear at β_{cr}, but for $\varepsilon \geqq \varepsilon^*(d)$ there is a jump, i.e. the corresponding scaling distributions are far from this branch.

Such a situation cannot be excluded but we have no indication of how it can be realized.

4. For the spontaneous breakdown of the continuous symmetry the appearance of scaling distributions is a very interesting problem. It is fairly probable that the translation-invariant limit Gibbs distributions in systems with broken symmetry belong to the domain of attraction of the scaling distributions if β is large.

These are the most interesting, yet unsolved, problems of the theory of phase transitions of the second kind.

Historical notes and references to Chapter IV

1. The definition of critical indices can be found in practically every book on statistical mechanics. We refer to Stanley's book [43] as an example.

2. Dyson's hierarchical models were introduced in his paper [59]. The analysis of these models from the point of view of critical point was first carried out on the physical level in Baker's paper [45]. These models were also mentioned by Wilson in his lectures (see [94]). The paper [47] of Bleher and Sinai contains a rigorous mathematical analysis of the Gaussian solution. The treatment given here on the whole follows [47], but a great deal of the technical details are essentially simplified. The case $c=\sqrt{2}$ was separately analysed by Bleher [5]. In Bleher's work [6] the neighbourhood of β_{cr} for a Gaussian point is investigated in detail.

3. The non-Gaussian solution of the basic integral equation for hierarchical models were constructed by Bleher and Sinai in [48]. In the same paper, the results of numerical computations were also given for the non-Gaussian branch made by Bleher. It would be very interesting to continue these investigations. It is probable that this is an example for which rigorous results could be obtained on the basis of computer experiments.

The first terms of the expansion for the critical indices with respect to the parameter $\varepsilon=\sqrt{2}-c$ are found in [48]. In the papers [83], [93] by Thompson et al., the values of these coefficients were more precisely determined. Interesting results were obtained by Collet and Eckmann [54], [119], in particular they proved that the critical indices are infinitely differentiable functions of ε. We also refer to McGuire's paper [101] on the spherical hierarchical models.

4. The renormalization group method in the theory of phase transitions was developed by Fisher, Kadanoff and Wilson. Since then, this has become one of the basic methods of the theory of critical points. A good presentation of its application by physicists can be obtained from the surveys of Kogut and Wilson [94], Fisher [60], Ma [99], Brezin, Le Guillou and Zinn-Justin [50], Patashinsky and Pokrovsky [31].

5. The notion of scaling distribution was introduced by Sinai [42], Gallavotti and Jona-Lasinio [70]. For the discussion of this notion see the papers of Gallavotti and Knops [69], di Castro and Jona-Lasinio [52], Cassandro and Gallavotti [53], Jona-Lasinio [88]. The continuous scaling fields were studied by Dobrushin in [18], [58]. The one-dimensional scaling field already appeared in Kolmogorov's work [22] on the problems of similarity in the theory of turbulence (see [32], [44], and also Sinai's lectures [112], [113]). The Gaussian scaling fields in the discrete case were constructed in Sinai's work [42]. The works of Bleher [7] and Dobrushin and Takahashi [20] contain examples for Gaussian scaling distributions with spectra which cannot be described by (4.15).

6. The general theory of Hermite–Ito polynomials can be found for example, in Ito's paper [90] Simon's book [41] and the extensive survey of Dobrushin and Minlos [19].

7. The spectrum of the linearized renormalization group for the Gaussian scaling distribution is studied in Bleher's work [7] approximately in the same spirit as in Sections 7 and 8. Many of the results of Sections 7 and 8, in particular Lemma 4.9, are due to Missarov (a student of the Moscow University).

8. The ε-expansions described in Section 9 appeared in the papers of Fisher, Ma and Nickel [61] and Zak [117] on the physical level of rigour. Non-Gaussian scaling distributions figure in Rosenblatt's paper [108]. The examples given by Dobrushin in papers [18], [58] are approximately based on the same idea. There are other interesting examples for non-Gaussian scaling distributions in the works of Karwowski and Streit [92], Molchanov and Sudarev [30]. The preprint of Bramson and Griffeath ("Renormalizing in three-dimensional voter model", Courant Institute, 1977) contains an interesting example of the limit theorems for Gaussian scaling fields.

EPILOGUE

We mention here some works from the extensive literature that has recently been devoted to the problems treated in this book.

In connection with the content of Chapter I and some problems of Chapter IV we refer to the thorough analysis of the limit Gibbs distributions for arbitrary real-valued spin systems. In particular, this concerns the systems with Hamiltonians of the form

$$\mathscr{H} = \sum a(x-y)\varphi(x)\varphi(y) + \sum b(x_1, x_2, x_3, x_4)\varphi(x_1)\varphi(x_2)\varphi(x_3)\varphi(x_4)$$

where some positivity conditions for the fourth order term are required. The basic problem is to investigate the character of the decay of the finite-dimensional distributions of the limit Gibbs states when $|\varphi(x)| \to \infty$. A great amount of results in this direction known by the author is obtained by J. Lebowitz and E. Presutti [97], and M. Cassandro, E. Olivieri, A. Pellegrinotti et al. (Zeitschr. für Wahrscheinlichs-keitstheorie, 1978, vol. 41, p. 313).

Chapter II is devoted to the analysis of the phase diagrams of lattice systems. An other method — different from the one presented in this book — of proving the non-uniqueness of limit Gibbs distributions is developed in a series of papers which are published in J. of Stat. Phys. by four authors J. Fröhlich, R. B. Israel, E. Lieb and B. Simon (FILS). The basic ideas of this method are, first, the reflection positivity which is close to the positive definiteness of the corresponding transfer matrix and, second, a special estimation of the probabilities of contours (chessboard method). In the paper of FILS published in Comm. Math. Phys. (1978, vol. 62, No. 1, p. 1) this method is compared with that of Chapter II in detail. We can agree with the conclusions of this comparison except for the last one that concerns the difficulty of the proofs. However, there are some indications that a great deal of results of FILS can be obtained by appropriate improvements of the methods of Chapter II. The basic example, which is natural to start with, could be the proof via the contour method of the non-uniqueness of the limit Gibbs distribution of the $:P(\varphi):_2$ quantum field theory.

Evidently, this requires a generalization of contour models for the case of interacting contours.

An other interesting direction is to extend the results of Chapter II for lattice systems where the spin variable takes continuous values. We refer to the work of V. A. Malyshev and Yu. A. Terlecky and that of Yu. A. Terlecky (DAN SSSR 1979, vol. 246, No. 3, p. 540). In a paper of J. Slawny (J. Stat. Phys. 1979, vol. 20, No. 5, p. 57) phase diagrams for many cases are investigated by means of the perturbation theory in the neighbourhood of $\beta=\infty$.

In connection with Section 1 of Chapter III we mention the paper of T. Spencer and O. Mc Brian (Comm. Math. Phys. 1977, vol. 53, No. 3, p. 235) and that of S. B. Shlosman (Teor. Mat. Fiz. 1978, vol. 37, No. 3 p. 427) on the decay of correlations in two-dimensional models.

One may expect a significant development in the investigations of the problem of Chapter IV.

Firstly this concerns the analysis of the domain of attraction of stable Gaussian scaling distributions. An other problem is the construction of non-Gaussian scaling distributions by means of bifurcation theory and ε-expansions, in an analogous way as it was done in [48] for Dyson's hierarchical models. A great progress has been achieved by P. M. Bleher and M. D. Missarov, who could — in some formal sense — sum up the series of the ε-expansion and obtain a nice, compact expression for the corresponding scaling distribution and for the spectrum of the linearized renormalization group.

On the other hand, for the time being it is not clear how these investigations help us to analyse the critical indices of short-range models.

For this it is very important to investigate the scaling distributions corresponding to the exactly solvable models — the Ising model and the Baxter models. Extremely strong methods were developed here by M. Sato and his co-workers. Seemingly the development of this investigations is one of the most promising directions in the theory of phase transitions.

REFERENCES

References 1–44 are published in Russian

1. V. I. ARNOLD (1972): Lectures on bifurcations and versal families. *Usp. Mat. Nauk,* **27**, No. *5*, 119–184
2. F. A. BEREZIN, YA. G. SINAI (1967): The existence of phase transition for a lattice gas with attraction between the particles. *Trudy Mosk. Mat. Obshch.* **17**, 197–212
3. V. L. BEREZINSKY (1972): *Low Temperature Properties of Two-dimensional Systems with a Continuous Group of Symmetry.* Thesis, Landau Institute of Theoretical Physics, Moscow
4. V. B. BERESTECKY (1973): Gauge symmetries and a unified theory of the weak and of the electromagnetic interactions. In: *Elementary Particles* (Institute of Theoretical and Experimental Physics). Atomizdat, Moscow, Vol. **1**, pp. 3–25
5. P. M. BLEHER (1977): On phase transitions of the second order in asymptotically hierarchical models of Dyson. *Usp. Mat. Nauk* **32**, No. *6*, 243–244
6. P. M. BLEHER (1976): Phase transition of second order in certain ferromagnetic models. *Trudy Mosk. Mat. Obshch.* **33**, 155–222
7. P. M. BLEHER (1978): ε-expansion for scaling random fields. In: *Multi-component stochastic systems.* Nauka, Moscow, pp. 47–83
8. N. N. BOGOLUBOV, D. YA. PETRINA, B. M. HACET (1969): Mathematical description of equilibrium states of classical systems on the basis of the formalism of canonical ensemble. *Teor. Mat. Fiz.* **1**, 251–274
9. N. N. BOGOLUBOV, B. M. HACET (1949): On some mathematical questions of the theory of statistical equilibrium. *Dokl. Akad. Nauk SSSR* **66**, 321–324
10. V. M. GERCIK, R. L. DOBRUSHIN (1974): Gibbs states in a lattice model with next nearest neighbour interaction. *Funkts. Anal. Prilozh.* **8**, 12–25
11. V. M. GERCIK (1976): Conditions for the non-uniqueness of Gibbs states in lattice models with finite interaction potentials. *Izv. Akad. Nauk SSSR Ser. Mat.* **40**, 448–462
12. R. L. DOBRUSHIN (1965): Existence of a phase transition in two- and three-dimensional lattice models. *Teor. Veroyatn. Primen.* **10**, 209–230
13. R. L. DOBRUSHIN (1968): Problem of uniqueness of a Gibbs random field and phase transitions. *Funkts. Anal. Prilozh.* **2**, 44–57
14. R. L. DOBRUSHIN (1968): The description of the random field by its conditional distributions and its regularity conditions. *Teor. Veroyatn. Primen.* **2**, 201–229
15. R. L. DOBRUSHIN (1969): Gibbs field: the general case. *Funkts. Anal. Prilozh.* **3**, 27–35
16. R. L. DOBRUSHIN (1970): Definition of random variables by conditional distributions. *Teor. Veroyatn. Primen.* **15**, 469–497
17. R. L. DOBRUSHIN (1972): Gibbsian state which describes co-existence of phases for a three-dimensional Ising-model. *Teor. Veroyatn. Primen.* **17**, 612–639

17a. R. L. Dobrushin (1973): An investigation of Gibbsian states for three-dimensional lattice systems. *Teor. Veroyatn. Primen.* **18**, 261–279

18. R. L. Dobrushin (1978): Scaling property and renormalization group of generalized random fields. In: *Multicomponent stochastic systems.* Nauka, Moscow, pp. 179–214

19. R. L. Dobrushin, R. A. Minlos (1977): Polynomials of linear random functions. *Usp. Mat. Nauk* **32**, No. *2*, 67–122

20. R. L. Dobrushin, H. Takahashi: Personal communication

21. I. Kashapov (1978): The structure of the ground states for the three-dimensional Ising model with three-nearest-neighbour interaction. *Teor. Mat. Fiz.* **33**, No. *1*, 110–116

22. A. N. Kolmogorov (1940): The Wiener spiral and some other interesting curves in the Hilbert space. *Dokl. Akad. Nauk SSSR* **26**, 115–118

23. D. G. Martirosyan (1975): To the problem of an upper bound for the numbers of periodical Gibbs states for lattice gas models. *Usp. Mat. Nauk.* **30**, No. *6*, 181–182

24. D. G. Martirosyan (1975): Uniqueness of the limit Gibbs distribution for the perturbed Ising model. *Teor. Mat. Fiz.* **22**, 335

25. R. A. Minlos (1967): Limit Gibbs distributions. *Funkts. Anal. Prilozh.* **1**, 60–73

26. R. A. Minlos (1967): Regularity of a limit Gibbs distribution. *Funkts. Anal. Prilozh.* **1**, 40–54

27. R. A. Minlos, Ya. G. Sinai (1967): New results on phase transitions of the first order in lattice gas models with attraction between particles. *Trudy Mosk. Mat. Obshch.* **17**, 213–242

28. R. A. Minlos, Ya. G. Sinai (1967): The phenomenon of "phase separation" at low temperature in some lattice gas models. I. *Mat. Sb.* **73**, 375–448

29. R. A. Minlos, Ya. G. Sinai (1968): The phenomenon of "phase separation" at low temperature in some lattice gas models. II. *Trudy Mosk. Mat. Obshch.* **18**, 113–178

30. S. A. Molchanov, Yu. N. Sudarev (1975): Gibbs states in a spherical model. *Dokl. Akad. Nauk SSSR* **224**, 536–539

31. A. Patashinsky, V. L. Pokrovsky (1975): *Fluctuational Theory of Phase Transitions.* Nauka, Moscow

32. M. S. Pinsker (1955): The theory of curves in Hilbert spaces with stationary increments of the nth order. *Izv. Akad. Nauk. SSSR Ser. Mat.* **19**, No. *5*, 319–345

33. S. A. Pirogov (1974): First-order phase transitions for spin systems with spin values $-1, 0, 1$. *Dokl. Akad. Nauk. SSSR.* **214**, No. *6*, 1273–1275

34. S. A. Pirogov (1975): Phase coexistence for lattice models with several types of particles. *Izv. Akad. Nauk. SSSR Ser. Mat.* **39**, 1404–1442

35. S. A. Pirogov, Ya. G. Sinai (1974): Phase transitions of the second order for small perturbations of the Ising model. *Funkts. Anal. Prilozh.* **8**, 25–31

36. S. A. Pirogov, Ya. G. Sinai (1975): Phase diagrams of classical lattice systems I. *Teor. Mat. Fiz.* **25**, 358–369

37. S. A. Pirogov, Ya. G. Sinai (1976): Phase diagrams of classical lattice systems. II. *Teor. Mat. Fiz.* **26**, 61–76

38. Yu. A. Rozanov (1967): On the Gaussian homogeneous fields with given conditional distributions. *Teor. Veroyatn. Primen.* **12**, 433–443

39. D. Ruelle (1971): *Statistical Mechanics. Rigorous Results.* Mir. [Original edition in English: Benjamin, New York, Amsterdam, 1969]

40. D. Ruelle (1977): On manifolds of phase coexistence. *Teor. Mat. Fiz.* **30**, 40–47

41. B. Simon (1976): $P(\Phi)_2$ *Euclidean Quantum Field Theory.* Mir. [Original edition in English: Princeton, 1974]

42. Ya. G. Sinai (1976): Automodel probability distributions. *Teor. Veroyatn. Primen.* **21**, 63–78

43. H. E. STANLEY (1973): *Phase Transitions and Critical Phenomena.* Mir. [Original edition in English: Oxford University Press, 1971]
44. A. M. YAGLOM (1957): Certain types of random fields in n-dimensional space similar to stationary stochastic processes. *Teor. Veroyatn. Primen.* **2**, 292–338

References 45–119 are given with their original titles

45. G. BAKER (1972): Ising model with a scaling interaction. *Phys. Rev.* **B5**, 2622–2633
46. H. VAN BEIJEREN (1975): Interface sharpness in the Ising system. *Comm. Math. Phys.* **40**, 1–7
47. P. M. BLEHER, YA. G. SINAI (1973): Investigation of the critical point in models of the type of Dyson's hierarchical models. *Comm. Math. Phys.* **33**, 23–42
48. P. M. BLEHER, YA. G. SINAI (1975): Critical indices for Dyson's asymptotically hierarchical models. *Comm. Math. Phys.* **45**, 247–278
49. A. BORTZ, R. GRIFFITHS (1972): Phase transitions in an isotropic classical Heisenberg ferromagnet. *Comm. Math. Phys.* **26**, 102–108
50. E. BREZIN, J. C. LE GUILLOU, J. ZINN-JUSTIN (1976): *Phase Transitions and Critical Phenomena* edited by Domb and Green, Vol. 6. Academic Press
51. M. CASSANDRO, A. DA FANO, E. OLIVIERI (1975): Existence of phase transition for a lattice model with a repulsive hard core and an attractive short range interaction. *Comm. Math. Phys.* **44**, 45–51
52. M. CASSANDRO, G. JONA-LASINIO: *Renormalization Group Approach to Critical Behaviour.* Preprint
53. M. CASSANDRO, G. GALLAVOTTI (1975): The Lavoisier law and the critical point. *Nuovo Cim.* **25B**, 691–705
54. P. COLLET, J. P. ECKMANN (1977): The ε-expansion for the hierarchical model. *Comm. Math. Phys.* **55**, 67–96
55. R. L. DOBRUSHIN (1967): Existence of phase transitions in models of a lattice gas. *Proc. of the Fifth Berkeley Symposium,* III. 73–87
56. R. L. DOBRUSHIN, S. B. SHLOSSMAN (1975): Absence of breakdown of continuous symmetry in two-dimensional models of statistical physics. *Comm. Math. Phys.* **42**, 31–40
57. R. L. DOBRUSHIN, YA. G. SINAI (1978): *Multicomponent Random Systems.* Marcel Dekker, New York
58. R. L. DOBRUSHIN: Gaussian and their subordinated selfsimilar random fields. *Ann. Prob.* (to appear)
59. F. DYSON (1969): Existence of a phase transition in a one-dimensional Ising ferromagnet. *Comm. Math. Phys.* **12**, 91–107
59a. F. DYSON (1971): An Ising ferromagnet with distributions of long-range order. *Comm. Math. Phys.* **21**, 269–283
60. M. E. FISHER (1974): The renormalization group in the theory of critical behaviour. *Rev. Mod. Phys.* **46**, 597–616
61. M. E. FISHER, S. K. MA, B. G. NICKEL (1972): Critical exponent for long-range interactions. *Phys. Rev. Lett.* **29**, 917–920
62. J. FELDMAN, K. OSTERWALDER (1976): The Wightman axioms and the mass gap for weakly coupled φ_3^4 quantum field theories. *Ann. Phys.* (NY) **97**, 80–135
63. J. FRÖHLICH (1975): Poetic phenomena in two-dimensional quantum field theory: non-uniqueness of the vacuum, the solitons and all that. In: *Les Méthodes Mathématiques de la Théorie Quantique des Champs,* Marseilles, pp. 111–130

64. J. FRÖHLICH, B. SIMON, T. SPENCER (1976): Infrared bounds, phase transitions and continuous symmetry breaking. *Comm. Math. Phys.* **50**, 79–85

65. J. FRÖHLICH (1976): Phase transitions, Goldstone bosons and topological sugar-selection rules. *Acta Phys. Austriaca*, Suppl. **15**, 133

66. J. FRÖHLICH (1978): The pure phases (harmonic functions of generalized processes) or: Mathematical physics of phase transitions and symmetry breaking. *Bull. Am. Math. Soc.* **84**, No. *2*, 165–193

67. J. FRÖHLICH (1977): *Quantum Sin-Gordon equation and quantum solitons in two space-time dimensions.* Lectures delivered at the International School of Mathematical Physics, "Ettore Majorane", Erice, Sicily

68. G. GALLAVOTTI, S. MIRACLE-SOLE (1972): Equilibrium states of the Ising model in the two phase region. *Phys. Rev.* **5B**, 2555

69. G. GALLAVOTTI, H. J. F. KNOPS (1974): Block-spin interactions in the Ising model. *Comm. Math. Phys.* **36**, 171–184

70. G. GALLAVOTTI, G. JONA-LASINIO (1975): Limit theorems for multidimensional Markov processes. *Comm. Math. Phys.* **41**, 301–307

71. K. GAWEDZKI (1978): Existence of three phases for $:P(\varphi):_2$ model of quantum field. *Comm. Math. Phys.* **59**, 117–142

72. J. GINIBRE, A. GROSSMAN, D. RUELLE (1966): Condensation of lattice gases. *Comm. Math. Phys.* **3**, 187

73. J GLIMM, A. JAFFE (1971): Quantum field models. In: *Statistical Mechanics and Quantum Field Theory,* Les Houches (ed. by C. De Witt and R. Store), Gordon and Breach, New York

74. J. GLIMM, A. JAFFE (1972): Boson quantum field models. In: *Mathematics of Contemporary Physics* (ed. by R. Streater) Academic Press, New York, pp. 77–145

75. J. GLIMM, A. JAFFE (1976): A tutorial course in constructive field theory. In: *Proc. of Cargese Summer School* (ed. by C. de Witt and R. Stora) Gordon & Breach, New York

76. J. GLIMM, A. JAFFE (1973): Positivity of the $(\Phi^4)_3$ Hamiltonian. *Fortschr. Phys.* **21**, 327–376

77. J. GLIMM, T. SPENCER (1975): Phase transitions for Φ_2^4 quantum field. *Comm. Math. Phys.* **45**, 203–216.

78. J. GLIMM, A. JAFFE, T. SPENCER (1976): A convergent expansion about mean field theory. *Ann. Phys.* (NY) **101**, 610–669.

79. R. GRIFFITHS (1964): A proof that the free energy of a spin system is extensive. *J. Math. Phys.* **5**, 1215–1222.

80. C. GRUBER, A. HINTERMANN, D. MERLINI (1977): *Group Analysis of Classical Lattice Systems,* Lecture Notes in Physics No. *60*, Springer, Berlin, Heidelberg, New York

81. F. GUERRA, L. ROSEN, B. SIMON (1975): The $P(\Phi)_2$ euclidean quantum field theory as classical statistical mechanics. *Ann. Math.* **101**, 111–259

82. F. GUERRA (1975): Local algebras in euclidean quantum field theory. In: Proceedings of the Conference, *C*-algebras and their Applications in Theoretical Physics,* Rome, pp. 13–26

83. A. J. GUTTMAN, D. KIM, C. J. THOMPSON (1977): Critical properties of Dyson's hierarhical model. *J. Phys. A. Math. Gen.*, **10**, No. *9*, 1579–1598

84. O. J. HEILMANN, E. H. LIEB (1970): Theory of monomer-dimer systems. *Phys. Rev. Lett.* **84**, 1412

84a O. J. HEILMANN, E. H. LIEB, O. J. LIEB (1972): Theory of monomer-dimer systems. *Comm. Math. Phys.* **25**, 190–232

85. O. J. HEILMANN (1974): The use of reflection as symmetry operation in connection with Peierls's argument. *Comm. Math. Phys.* **36**, 91–114

86. W. Holsztynski, J. Slawny (1976): *Phase Transitions in Ferromagnetic Systems at Low Temperatures.* Preprint
87. W. Holsztynski, J. Slawny (1978): Peierls condition and number of ground states. *Comm. Math. Phys.* **61**, 177–190
88. G. Jona-Lasinio (1975): The renormalization group: a probabilistic view. *Nuovo Cim.* **26B**, 99–119
89. R. B. Israel (1975): Existence of phase transitions for long-range interactions. *Comm. Math. Phys.* **43**, 59–68
90. K. Ito (1951): Multiple Wiener integral. *J. Math. Soc. J.*, **3**, 157–169
91. L. Kadanoff et al. (1967): *Rev. Mod. Phys.* **39**, 395–431
92. W. Karwowski, L. Streit (1976): *A Renormalization Group Model with Non-Gaussian Fixed Point,* Universität Bielefeld, Preprint.
93. D. Kim, C. J. Thompson (1977): Critical properties of Dyson's hierarchical model. *J. Phys. A: Math. Gen.* **10**, 1579–1598
94. J. Kogut, K. Wilson (1974): The renormalization group. *Phys. Rep.*, **12C**, 75
95. O. E. Lanford, D. Ruelle (1969): Observables at infinity and states with short-range correlations in statistical mechanics. *Comm. Math. Phys.* **13**, 194–215
96. J. Lebowitz (1977): Coexistence of phases in Ising ferromagnets, *J. Stat. Phys.* **16**, 463–477
97. J. Lebowitz, E. Presutti (1976): Statistical mechanics of systems of unbounded spins. *Comm. Math. Phys.* **50**, 195–218
98. E. H. Lieb (1978): New proofs of long-range order. In: *Mathematical Problems in Theoretical Physics,* Proceedings, Rome, 1977. Lecture Notes in Physics, No. *80*, pp. 59–67. Springer, Berlin, Heidelberg, New York
99. S. K. Ma (1973): Introduction to the renormalization group. *Rev. Mod. Phys.* **45**, 589
100. V. A. Malyshev (1975): Phase transitions in classical Heisenberg ferromagnets with arbitrary parameter of anisotropy. *Comm. Math. Phys.* **40**, 75–82
101. J. B. McGuire (1973): The spherical hierarchical model. *Comm. Math. Phys.* **32**, 215–230
102. A. Messager, S. Miracle-Sole (1975): Equilibrium states of the two-dimensional Ising model in the two-phase region. *Comm. Math. Phys.* **40**, 187–196
103. E. Nelson (1973): Probability theory and Euclidean field theory. In: *Constructive Quantum Field Theory* (ed: by G. Velo and A. Wightman). Lecture Notes in Physics, No. *25*, Springer, Berlin, Heidelberg, New York, pp. 94–124
104. K. Osterwalder, E. Seiler (1978): Gauge fields theories on the lattice. *Ann. Phys.* (NY) **11**, 440–471
105. R. Peierls (1936): On Ising's model of ferromagnetism. *Proc. Cambridge Philos. Soc.* **36**, 477–481
106. S. A. Pirogov, Ya. G. Sinai (1977): Ground states in two-dimensional boson quantum field theory. *Ann. Phys.* (NY) **109**, 393–400
107. C. H. Preston (1976): *Random Fields,* Lecture Notes in Mathematics. No. *534*, Springer, Berlin, Heidelberg, New York
108. M. Rosenblatt (1961): Independence and dependence. In: *Proceedings of the Fourth Berkeley Symposium on Probability and Mathematical Statistics,* Berkeley, pp. 411–443
109. D. Ruelle (1963): Classical statistical mechanics of a system of particles. *Helv. Phys. Acta,* **36**, 183–197
110. D. Ruelle (1977): A heuristic theory of phase transitions. *Comm. Math. Phys.* **53**, 195–208
111. L. K. Runnels (1975): Phase transitions of hard sphere lattice gases. *Comm. Math. Phys.* **40** 1–37

112. Ya. G. Sinai (1976): Some rigorous results in the theory of phase transitions. In: *Proceedings of the International Conference on Statistical Physics,* Budapest, 1975, pp. 139–144
113. Ya. G. Sinai (1978): Mathematical foundations of the renormalization group method. In: *Mathematical Problems in Theoretical Physics,* Proceedings 1977. Lecture Notes in Physics, No. *80*, Springer, Berlin, Heidelberg, New York, pp. 303–311
114. J. Slawny (1974): A family of equilibrium states relevant to low temperature behaviour of spin 1/2 classical ferromagnets. Breaking of translation symmetry. *Comm. Math. Phys.* **35**, 297–305
115. F. Spitzer (1974): Introduction aux processus de Markov à paramètre dans Z_r. In: *École d'Été de Probabilité de Saint-Flour.* III, 1973. Lecture Notes in Mathematics, No. *390*, Springer, Berlin, Heidelberg, New York
116. K. Wilson (1974): Confinement of quarks. *Phys. Rev.* **D 10**, 2445–2459
117. J. Zak (1973): Recursion relations and fixed points for ferromagnets with long-range interactions. *Phys. Rev.* **B 8**, 281–285
118. H. Bateman, A. Erdélyi (1953–55): *Higher Transcendental Functions,* Vols. 1–3, McGraw-Hill, New York
119. P. Collet, J.-P. Eckmann (1978): *A Renormalization Group Analysis of the Hierarchical Model in Statistical Mechanics.* Lecture Notes in Physics, No. *74*, Springer, Berlin, Heidelberg, New York

INDEX

OTHER TITLES IN THE SERIES IN NATURAL PHILOSOPHY

Vol. 1. Davydov—Quantum Mechanics (2nd Edition)
Vol. 2. Fokker—Time and Space, Weight and Inertia
Vol. 3. Kaplan—Interstellar Gas Dynamics
Vol. 4. Abrikosov, Gor'kov and Dzyaloshinskii—Quantum Field Theoretical Methods in Statistical Physics
Vol. 5. Okun'—Weak Interaction of Elementary Particles
Vol. 6. Shklovskii—Physics of the Solar Corona
Vol. 7. Akhiezer *et al.*—Collective Oscillations in a Plasma
Vol. 8. Kirzhnits—Field Theoretical Methods in Many-body Systems
Vol. 9. Klimontovich—The Statistical Theory of Nonequilibrium Processes in a Plasma
Vol. 10. Kurth—Introduction to Stellar Statistics
Vol. 11. Chalmers—Atmospheric Electricity (2nd Edition)
Vol. 12. Renner—Current Algebras and their Applications
Vol. 13. Fain and Khanin—Quantum Electronics, Volume 1—Basic Amplifiers and Oscillators
Vol. 14. Fain and Khanin—Quantum Electronics, Volume 2—Maser Theory
Vol. 15. March—Liquid Metals
Vol. 16. Hori—Spectral Properties of Disordered Chains and Lattices
Vol. 17. Saint James, Thomas and Sarma—Type II Superconductivity
Vol. 18. Margenau and Kestner—Theory of Intermolecular Forces (2nd Edition)
Vol. 19. Jancel—Foundations of Classical and Quantum Statistical Mechanics
Vol. 20. Takahashi—An Introduction to Field Quantization
Vol. 21. Yvon—Correlations and Entropy in Classical Statistical Mechanics
Vol. 22. Penrose—Foundations of Statistical Mechanics
Vol. 23. Visconti—Quantum Field Theory, Volume 1
Vol. 24. Furth—Fundamental Principles of Theoretical Physics
Vol. 25. Zheleznyakov—Radioemission of the Sun and Planets
Vol. 26. Grindlay—An Introduction to the Phenomenological Theory of Ferroelectricity
Vol. 27. Unger—Introduction to Quantum Electronics
Vol. 28. Koga—Introduction to Kinetic Theory: Stochastic Processes in Gaseous Systems
Vol. 29. Galasiewicz—Superconductivity and Quantum Fluids
Vol. 30. Constantinescu and Magyari—Problems in Quantum Mechanics
Vol. 31. Kotkin and Serbo—Collection of Problems in Classical Mechanics
Vol. 32. Panchev—Random Functions and Turbulence
Vol. 33. Taipe—Theory of Experiments in Paramagnetic Resonance
Vol. 34. Ter Haar—Elements of Hamiltonian Mechanics (2nd Edition)
Vol. 35. Clarke and Grainger—Polarized Light and Optical Measurement

Vol. 36. Haug—Theoretical Solid State Physics, Volume 1
Vol. 37. Jordan and Beer—The Expanding Earth
Vol. 38. Todorov—Analytical Properties of Feynman Diagrams in Quantum Field Theory
Vol. 39. Sitenko—Lectures in Scattering Theory
Vol. 40. Sobel'man—Introduction to the Theory of Atomic Spectra
Vol. 41. Armstrong and Nicholls—Emission, Absorptiòn and Transfer of Radiation in Heated Atmospheres
Vol. 42. Brush—Kinetic Theory, Volume 3
Vol. 43. Bogolyubov—A Method for Studying Model Hamiltonians
Vol. 44. Tsytovich—An Introduction to the Theory of Plasma Turbulence
Vol. 45. Pathria—Statistical Mechanics
Vol. 46. Haug—Theoretical Solid State Physics, Volume 2
Vol. 47. Nieto—The Titius-Bode Law of Planetary Distances: Its History and Theory
Vol. 48. Wagner—Introduction to the Theory of Magnetism
Vol. 49. Irvine—Nuclear Structure Theory
Vol. 50. Strohmeier—Variable Stars
Vol. 51. Batten—Binary and Multiple Systems of Stars
Vol. 52. Rousseau and Mathieu—Problems in Optics
Vol. 53. Bowler—Nuclear Physics
Vol. 54. Pomraning—The Equations of Radiation Hydrodynamics
Vol. 55. Belinfante—A Survey of Hidden Variables Theories
Vol. 56. Scheibe—The Logical Analysis of Quantum Mechanics
Vol. 57. Robinson—Macroscopic Electromagnetism
Vol. 58. Gombas and Kisdi—Wave Mechanics and its Applications
Vol. 59. Kaplan and Tsytovich—Plasma Astrophysics
Vol. 60. Kovacs and Zsoldos—Dislocations and Plastic Deformation
Vol. 61. Auvray and Fourrier—Problems in Electronics
Vol. 62. Mathieu—Optics
Vol. 63. Atwater—Introduction to General Relativity
Vol. 64. Muller—Quantum Mechanics: A Physical World Picture
Vol. 65. Bilenky—Introduction to Feynman Diagrams
Vol. 66. Vodar and Romand—Some Aspects of Vacuum Ultraviolet Radiation Physics
Vol. 67. Willett—Gas Lasers: Population Inversion Mechanisms
Vol. 68. Akhiezer *et al.*—Plasma Electrodynamics, Volume 1—Linear Theory
Vol. 69. Glasby—The Nebular Variables
Vol. 70. Bialynicki-Birula—Quantum Electrodynamics
Vol. 71. Karpman—Non-linear Waves in Dispersive Media
Vol. 72. Cracknell—Magnetism in Crystalline Materials
Vol. 73. Pathria—The Theory of Relativity
Vol. 74. Sitenko and Tartakovskii—Lectures on the Theory of the Nucleus
Vol. 75. Belinfante—Measurement and Time Reversal in Objective Quantum Theory
Vol. 76. Sobolev—Light Scattering in Planetary Atmospheres
Vol. 77. Novakovic—The Pseudo-spin Method in Magnetism and Ferroelectricity
Vol. 78. Novozhilov—Introduction to Elementary Particle Theory
Vol. 79. Busch and Schade—Lectures on Solid State Physics
Vol. 80. Akhiezer *et al.*—Plasma Electrodynamics, Volume 2
Vol. 81. Soloviev—Theory of Complex Nuclei
Vol. 82. Taylor—Mechanics: Classical and Quantum

Vol. 83. Srinivasan and Parthasathy—Some Statistical Applications in X-Ray Crystallography
Vol. 84. Rogers—A Short Course in Cloud Physics
Vol. 85. Ainsworth—Mechanisms of Speech Recognition
Vol. 86. Bowler—Gravitation and Relativity
Vol. 87. Klinger—Problems of Linear Electron (Polaron) Transport Theory in Semiconductors
Vol. 88. Weiland and Wilhelmson—Coherent Non-Linear Interaction of Waves in Plasmas
Vol. 89. Pacholczyk—Radio Galaxies
Vol. 90. Elgaroy—Solar Noise Storms
Vol. 91. Heine—Group Theory in Quantum Mechanics
Vol. 92. Ter Haar—Lectures on Selected Topics in Statistical Mechanics
Vol. 93. Bass and Fuks—Wave Scattering from Statistically Rough Surfaces
Vol. 94. Cherrington—Gaseous Electronics and Gas Lasers
Vol. 95. Sahade and Wood—Interacting Binary Stars
Vol. 96. Rogers—A Short Course in Cloud Physics (2nd Edition)
Vol. 97. Reddish—Stellar Formation
Vol. 98. Patashinskii and Pokrovskii—Fluctuation Theory of Phase Transitions
Vol. 99. Ginzburg—Theoretical Physics and Astrophysics
Vol. 100. Constantinescu—Distributions and their Applications in Physics
Vol. 101. Gurzadyan—Flare Stars
Vol. 102. Lominadze—Cyclotron Waves in Plasma
Vol. 103. Alkemade—Metal Vapours in Flames
Vol. 105. Klimontovich—Kinetic Theory of Non-Ideal Gases and Non-Ideal Plasmas
Vol. 107. Sitenko—Fluctuations & Nonlinear Interactions of Wave in Plasma
Vol. 108. Sinai—Theory of Phase Transitions: Rigorous Results
Vol. 109. Davydov—Biology and Quantum Mechanics
Vol. 110. Demianski—Relativistic Astrophysics